© NASA, ESA

Welcome

Our Solar System is a truly amazing place about which we know so much, yet at the same time still have so much to learn. In Understanding The Solar System, we'll explore the inner workings of our fascinating cosmic neighbourhood, from the wonders of our own planet to the secrets of the Ice Giants and all of our planetary neighbours. Marvel at the incredible star that makes it all possible, and delve beneath the surface of Earth's very own natural satellite and some of the Solar System's strangest moons. We'll also bring you a host of articles on other fascinating space phenomena, such as comets, asteroids, alien storms and more.

UNDERSTANDING THE SOLAR SYSTEM

Contents

06 25 unbelievable facts about the Solar System
Why our neighbourhood could be the strangest place in the cosmos

THE SUN AND PLANETS

18 The Sun
The Solar System would be nothing if it weren't for the power of our nearest star

22 The swelling Sun
Scientists get a gruesome look at how our Sun will eat planets

24 Mercury
This minute world is arguably the least explored of the four terrestrial planets

28 22 things you didn't know about Venus
Earth's sister planet is an intriguing and mysterious world

34 What are planets like on the inside?
Even among the worlds of our Solar System we see a huge variety of planets

36 Earth
The rocky world that we call home is full of wonders

40 Complete guide to Mars
We're learning more about the Red Planet every day

48 How the planets would look...
... if they were at the same distance from Earth as the Moon

50 Jupiter
The largest planet has a lot to tell us, and Juno is on the case

54 Everything you need to know about Saturn
There's more to this gas giant than meets the eye

60 Secrets of the Ice Giants
Join us as we peek into the unknown

114

28

06

4

CONTENTS

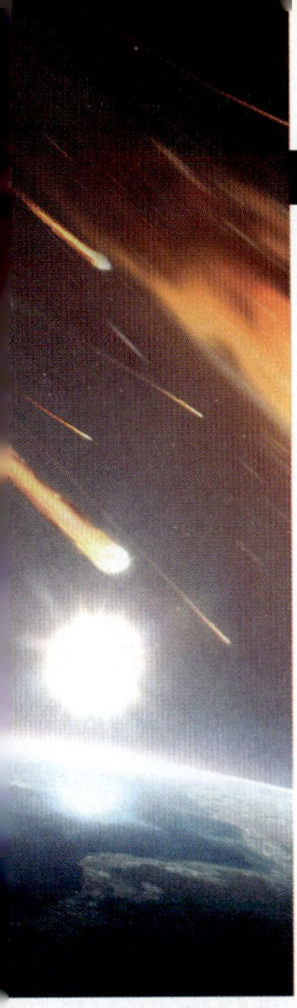

40

70

122

104

MOONS OF THE SOLAR SYSTEM

70 Guide to the Moon
Everything you need to know about our natural satellite

82 Strangest moons
Discover some of the fascinating worlds in the Solar System

88 Does Earth have a second moon?
Learn more about the asteroid tracking our orbit around the Sun

96 Escape to Titan
When the Sun scorches Earth, a tiny Saturnian moon could be our next home

OTHER SOLAR SYSTEM PHENOMENA

104 Space volcanoes
From Venus to Mars, volcanoes have helped shape the bodies of our Solar System

112 Martian megatsunami
How a megatsunami swept over the Red Planet

114 Comets, asteroids and meteor showers
Discover the space rocks that litter our Solar System

120 Where are the biggest craters?
Explore some of the largest craters in the Solar System

122 Alien storms
Discover incredible weather on other worlds and what causes it

5

UNDERSTANDING THE SOLAR SYSTEM

25 unbelievable facts about the Solar System

Why our neighbourhood could be the strangest place in the cosmos

25 UNBELIEVABLE FACTS ABOUT THE SOLAR SYSTEM

1 The Solar System is really big

NASA's Voyager 1 spacecraft was launched in 1977. More than three decades later, in 2012 it became the first human-made object to enter interstellar space by crossing the heliopause – the edge of the heliosphere. That's the boundary beyond which most of the Sun's ejected particles and magnetic fields dissipate. "If we define our Solar System as the Sun and everything that primarily orbits the Sun, Voyager 1 will remain within the confines of the Solar System until it emerges from the Oort Cloud in another 14,000 to 28,000 years," NASA says.

> "The Moon is both mind-bogglingly distant and incredibly close depending on how you think about it"

2 Even just our neighbourhood is really big

Depending on how carefully you do the calculations and how you arrange them, all of the planets in the Solar System could fit in between Earth and its Moon. The distance between Earth and the Moon varies as it orbits around us, as does the diameter of each of the planets – they're wider at their equators, so Saturn and Jupiter would have to be tilted sideways for this to work. But imagine lining them all up, pole to pole. They'd just barely squeeze in between us and our closest companion in space, blocking out the sky with their rings and gas giant bulk as they did so.

The Moon is the farthest from Earth we've ever sent humans, and it's both mind-bogglingly distant and incredibly close depending on how you think about it. Eight enormous planets could fit between here and there, and the distance from Earth to the Sun is more than 390 times the distance from Earth to the Moon. Scientists use an approximation of the Earth-Sun distance, also known as one astronomical unit, or AU, to compare distances within the Solar System. Jupiter is about 5.2 AU from the Sun, and Neptune is 30.07 AU from the Sun – around 30 times as far from the star as Earth.

3 The Sun's atmosphere is hotter than its surface

While the Sun's visible surface, the photosphere, is 5,500 degrees Celsius (10,000 degrees Fahrenheit), its upper atmosphere has temperatures in the millions of degrees. It's a large temperature differential with little explanation. NASA has several Sun-gazing spacecraft on the case, however, and they have some ideas for how the heat is generated. One is 'heat bombs', which happen when magnetic fields cross and realign in the corona. Another is when plasma waves move from the Sun's surface into the corona. With new data from the Parker Solar Probe – which recently became the first human-made object to 'touch' the Sun – coming in all the time, we're closer than ever to unlocking the mysteries at the heart of our Solar System.

UNDERSTANDING THE SOLAR SYSTEM

5. Venus is swept by powerful winds that could harbour life

Venus is a hellish planet with a high-temperature, high-pressure environment on its surface. Bone-dry and hot enough to melt lead, it's not exactly a welcoming environment, and has probably always been inhospitable to life. When the heavily shielded Venera spacecraft from the Soviet Union landed there in the 1970s, each lasted a few minutes, or at most a few hours, before melting or being crushed beyond their ability to function.

But even above its surface, the planet has a bizarre environment. Scientists have found that its upper winds flow 50 times faster than the planet's rotation. The European Venus Express spacecraft, which orbited the planet between 2006 and 2015, tracked the winds over long periods and detected periodic variations. It also found that the hurricane-force winds appeared to be getting stronger over time. A 2020 study that thrilled many astrobiologists detected phosphine, a possible sign of decaying biological matter, high in the Venusian clouds. Could they be a sign of life? Not without sufficient water, claim follow-up studies that reject the possibility of life in Venus' dry and windy atmosphere.

4. Mercury is still shrinking

Mercury is already the smallest planet in the Solar System, and is the second-densest after Earth. And it's only getting smaller and denser. For many years, scientists believed that Earth was the only tectonically active planet in the Solar System. But that changed after the Mercury Surface, Space Environment, Geochemistry and Ranging (MESSENGER) spacecraft did the first orbital mission at Mercury, mapping the entire planet in high definition and getting a look at the features on its cratered surface.

In 2016, data from MESSENGER revealed cliff-like landforms known as fault scarps. Because the fault scarps are relatively small, scientists can be sure that they weren't created that long ago and that the small planet is still contracting 4.5 billion years after the Solar System was formed.

6. Earth's Van Allen belts are more bizarre than expected

There are several bands of magnetically trapped, highly energetic charged particles surrounding our planet, known as the Van Allen belts. While we've known about the belts since the dawn of the Space Age, the Van Allen Probes, launched in 2012, have provided our best ever view of them. They've uncovered quite a few surprises along the way. We now know that the belts expand and contract according to solar activity. Sometimes the belts are very distinct from one another, and sometimes they swell into one massive unit. An extra radiation belt beyond the known two was spotted in 2013. Understanding these belts helps scientists make better predictions about space weather or solar storms.

25 UNBELIEVABLE FACTS ABOUT THE SOLAR SYSTEM

7 Organic molecules are everywhere

Organics are complex carbon-based molecules found in living things, but can be created by non-biological processes too. While organic molecules are common on Earth, they can be found in many other places in the Solar System too. Scientists have found organics on the surface of Comet 67P/Churyumov–Gerasimenko, for example. The discovery bolstered the case that organic molecules on Earth could have been brought to the surface from space. Organics have also been found on the surface of Mercury, on Saturn's moon Titan – giving Titan its orange colour – and on Mars.

"While organic molecules are common on Earth, they can be found in many other places too"

8 A valley on Mars could swallow the Grand Canyon

At 4,000 kilometres (2,500 miles), the immense system of Martian canyons known as Valles Marineris is more than ten times as long as the Grand Canyon on Earth. Valles Marineris escaped the notice of earlier Mars spacecraft flying over other parts of the planet and was finally spotted by the global mapping mission Mariner 9 in 1971. And what a sight it was to miss – Valles Marineris could stretch from New York to Seattle. The lack of active plate tectonics on Mars makes it tough to figure out how the canyon formed. Some scientists think that a chain of volcanoes on the other side of the planet, known as Tharsis Ridge, which includes Olympus Mons, somehow bent the crust from the opposite side of Mars. That cataclysmic force activated cracks in the crust, vast amounts of subsurface water that emerged to carve away rock and glaciers that crunched new pathways into the canyon system.

9 Mars' biggest volcano is bigger than Hawaii

While Mars seems quiet now, gigantic volcanoes once dominated the surface of the planet. This includes Olympus Mons, the biggest volcano ever discovered in the Solar System. At 602 kilometres (374 miles) across, the volcano is comparable to the size of Arizona. It's 25 kilometres (16 miles) high, or triple the height of Mount Everest, the tallest mountain on Earth. By volume, Olympus Mons is 100 times larger than Earth's largest volcano, Hawaii's Mauna Loa. Scientists speculate that volcanoes on Mars can grow to such immense sizes because gravity there is much weaker than it is on Earth. In addition, while Earth's crust constantly moves, the Martian crust likely doesn't, although the debate among researchers continues. The Hawaiian islands were formed as a hotspot in the mantle created a chain of volcanoes in the crust cruising by above it, so if the surface of Mars isn't moving, a volcano could build up for longer in one spot.

UNDERSTANDING THE SOLAR SYSTEM

10 The Great Red Spot is shrinking

Along with being the Solar System's largest planet, Jupiter also hosts the Solar System's largest storm. Known as the Great Red Spot, it's been observed in telescopes since the 1600s and studied by modern instruments like those on NASA's Juno, which recently provided evidence that the storm is hundreds of miles tall and likely fed by winds from thousands of miles below. The storm has been a raging conundrum for centuries, but in recent decades another mystery emerged: the spot is getting smaller. In 2014 the storm was only 16,500 kilometres (10,250 miles) across, about half its historic size. The shrinkage is being monitored in professional telescopes, and also by amateurs. Amateurs are often able to make more consistent measurements of Jupiter because viewing time on larger, professional telescopes is limited and often split between different objects.

11 Jupiter's moon Io has huge volcanic eruptions

Compared to Earth's peaceful Moon, Jupiter's moon Io may come as a surprise. The Jovian moon has hundreds of volcanoes and is considered the most active moon in the Solar System, sending plumes of sulphur up to 300 kilometres (190 miles) into its atmosphere. As such, Io's volcanoes emit around one tonne of gases and particles into space near Jupiter each second. Io's eruptive nature is caused by the immense forces the moon is exposed to nestled in Jupiter's gravitational well, as well as its magnetic field. The moon's insides tense up and relax as it orbits closer to and farther from the planet, generating enough energy for volcanic activity. Scientists are still trying to figure out how heat spreads through Io's interior, though, making it difficult to predict where the volcanoes exist using scientific models alone.

"*The storm has been a raging conundrum for centuries, but in recent decades another mystery emerged: It's getting smaller*"

25 UNBELIEVABLE FACTS ABOUT THE SOLAR SYSTEM

12 There is water everywhere

Water was once considered rare in space. But water ice exists all over the Solar System. It's a common component of comets and asteroids. Water can be found as ice in permanently shadowed craters on Mercury and the Moon, though we don't know if there's enough to support prospective human colonies in those places. Mars also has ice at its poles, in frost and likely below the surface dust. Even smaller bodies in the Solar System have ice: Saturn's moon Enceladus and the dwarf planet Ceres, among others.

Scientists suspect Jupiter's moon Europa may be the most likely candidate for extraterrestrial life because against all expectations there's likely liquid water below its cracked and frozen surface. Europa, much smaller than Earth, may host a deep ocean that researchers suggest could contain twice as much water as all of Earth's oceans combined.

But we know that not all ice is the same. A close-up examination of Comet 67P by the European Space Agency's Rosetta spacecraft revealed a different kind of water ice than the kind found on Earth.

13 Saturn has a yin-yang moon

Iapetus has a very dark hemisphere that always faces away from the planet and a very light hemisphere that always faces towards Saturn. Most asteroids, moons and planets are relatively uniform across their surfaces, but Iapetus sometimes shines brightly enough to be spotted by telescopes and then dims down by several magnitudes when oriented in the other direction. Current research suggests that Iapetus is made mostly of water ice. As the moon's darker side faces the Sun, scientists hypothesise, water ice sublimates away from that area, leaving darker rock behind. That may have created a positive feedback loop, as dark material heats up more than bright, reflective ice – as the darker, warmer side of the moon loses its ice, it becomes easier to heat up each time it faces the Sun, hastening the loss of more ice.

14 Titan has a liquid cycle

Another weird moon in Saturn's system is Titan, which hosts a liquid 'cycle' that moves material between the atmosphere and the surface. That sounds a lot like Earth's water cycle, but Titan's immense lakes are filled with liquid methane and ethane, possibly over a layer of water. Researchers hope to use data from the Cassini-Huygens mission to tease out some of Titan's secrets before designing a submarine that might one day plumb the depths of the mysterious moon.

15 Saturn has a giant storm

Saturn's northern hemisphere features a raging six-sided storm nicknamed 'the hexagon'. It has been present on the ringed planet for decades, if not longer. It was discovered in the 1980s, but was barely visible until the Cassini mission flew by between 2004 and 2017. Images and data from Cassini reveal the storm to be 300 kilometres (180 miles) tall, 32,000 kilometres (20,000 miles) wide and composed of air moving at about 320 kilometres (200 miles) per hour.

16 Rings are more common than we thought

We've known about Saturn's rings since telescopes were invented in the 1600s, but it took spacecraft and more powerful telescopes built in the last 50 years to reveal more. We now know that every planet in the outer Solar System – Jupiter, Saturn, Uranus and Neptune – has a ring system. But the rings differ from planet to planet: Saturn's spectacular halo, made in part of sparkly, reflective water ice, is not repeated anywhere else. Instead the rings of the other giants are likely made of rocky particles and dust. Rings aren't limited to planets, either. In 2014 astronomers discovered rings around the asteroid 10199 Chariklo.

UNDERSTANDING THE SOLAR SYSTEM

17 Spacecraft have visited every planet

We've been exploring space for more than 60 years, and have been lucky enough to get close-up pictures of dozens of celestial objects. Most notably, we've sent spacecraft to all of the planets in our Solar System – Mercury, Venus, Earth, Mars, Jupiter, Saturn, Uranus and Neptune – as well as two dwarf planets, Pluto and Ceres. The bulk of the flybys came from NASA's Voyager 1 and Voyager 2, which left Earth more than four decades ago and are still transmitting data from interstellar space. Between them the Voyagers clocked visits to Jupiter, Saturn, Uranus and Neptune thanks to an opportune alignment of the outer planets.

"Between them the Voyagers clocked visits to Jupiter, Saturn, Uranus and Neptune"

18 Worlds could be contaminated by spacecraft

So far, scientists have found no evidence that life exists elsewhere in the Solar System. But as we learn more about how 'extreme' microbes live in underwater volcanic vents or frozen environments, more possibilities open up for where they could live on other planets.

Microbial life is now considered likely enough on Mars that scientists take special precautions to sterilise spacecraft headed to the planet. NASA chose to crash its Galileo spacecraft into Jupiter rather than risk it contaminating the potentially habitable oceans of Europa.

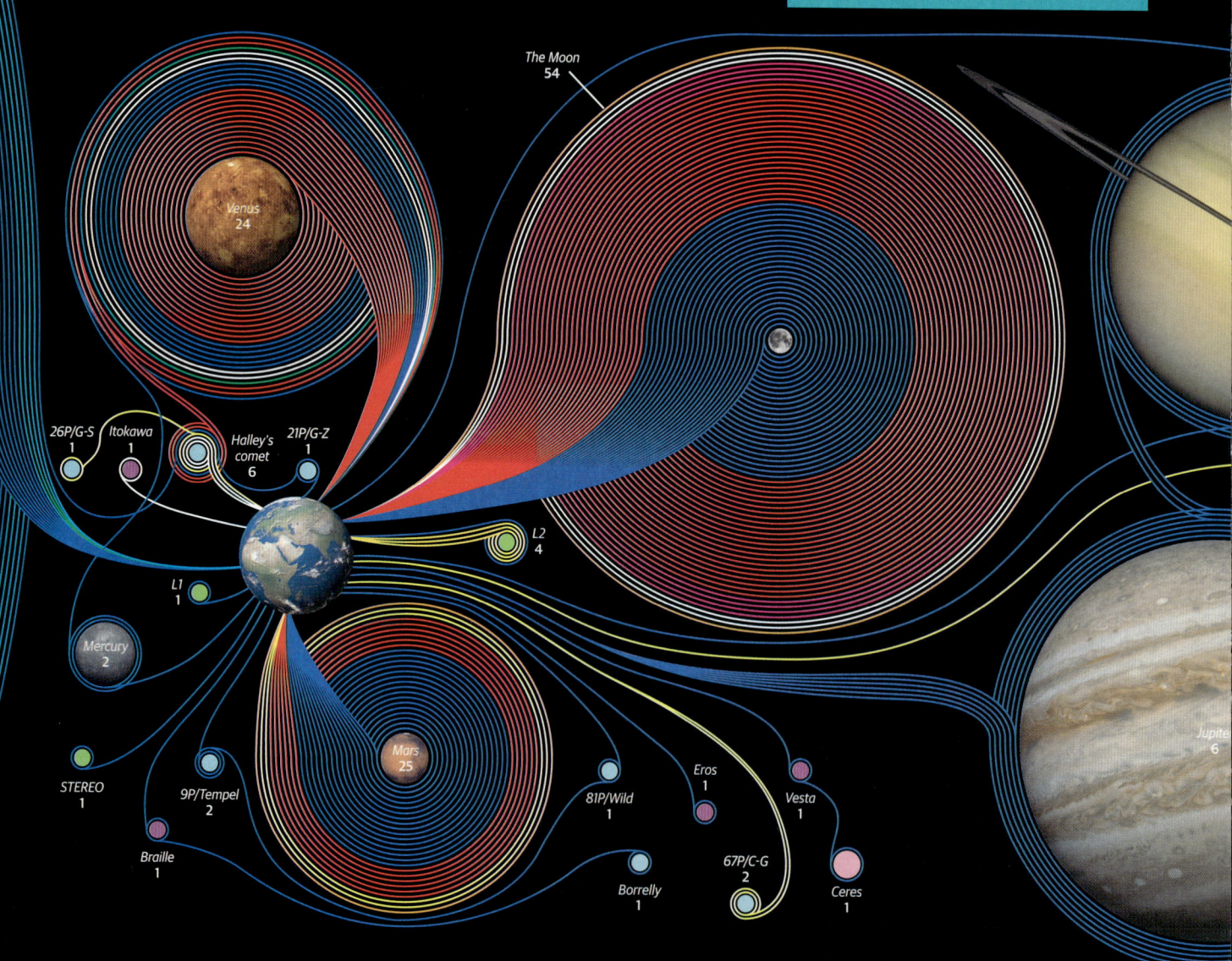

25 UNBELIEVABLE FACTS ABOUT THE SOLAR SYSTEM

19 Most comets are spotted with a Sungrazing telescope

Comets used to be the province of amateur astronomers who spent night after night scouring the skies with telescopes. While some professional observatories also made discoveries while viewing comets, that began to change with the launch of the Solar and Heliospheric Observatory (SOHO) in 1995. Since then, the spacecraft has found more than 2,400 comets, which is a pretty productive side mission for a telescope meant to observe the Sun. These comets are nicknamed 'sungrazers'. Many amateurs still participate in the search for comets by picking them out from raw SOHO images. One of SOHO's most famous observations came when it watched the breakup of the bright Comet ISON in 2013.

20 Uranus spins sideways

This gas giant is pretty weird on closer inspection. First, the planet rotates on its side, appearing to roll around the Sun like a ball. The most likely explanation for the planet's unusual orientation – about 90 degrees sideways compared to the other planets – is that it underwent some sort of titanic collision in the ancient past. Uranus' tilt causes what NASA considers to be the most extreme seasons in the Solar System. For about a quarter of each Uranus year – or 21 Earth years, as each Uranus year is 84 years long – the Sun shines directly over the north or south pole of the planet. That means for more than two decades on Earth, half of Uranus never sees the Sun at all.

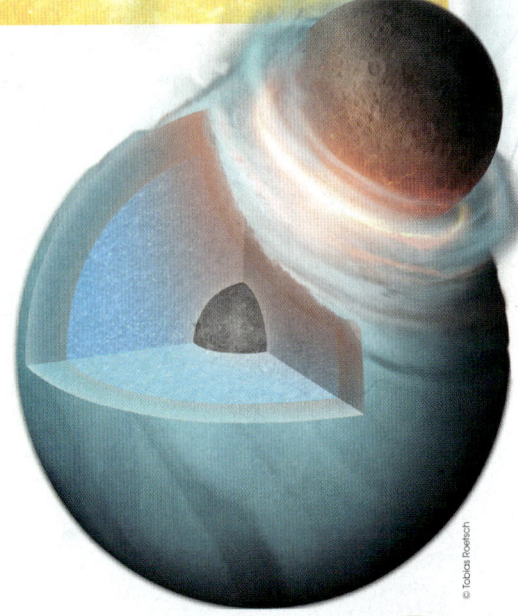

21 What happened to Miranda?

One of the most bizarre moons in the outer Solar System is Miranda, a shadowy moon of Uranus observed only once when Voyager 2 got a glimpse in 1986. Miranda hosts sharp ridges, craters and other major disruptions on its surface that would usually be the result of volcanic action. Tectonic activity could cause that kind of surface, but Miranda is much too small to generate that kind of heat on its own. Researchers think that gravitational pull from Uranus could have generated the push-pull action needed to heat, churn and contort Miranda's surface. To know for sure, we'll need to send another spacecraft to check out the moon's unobserved northern hemisphere.

UNDERSTANDING THE SOLAR SYSTEM

22 Neptune is too hot

Neptune is roughly 30 times as far from the Sun as Earth, and it gets correspondingly less heat and light. But it radiates far more heat than it's taking in and has far more activity in its atmosphere than planetary scientists would suspect, especially compared to nearby Uranus. Uranus is closer to the Sun and yet radiates about the same amount of heat as Neptune, and scientists aren't sure why. Winds on Neptune can blow up to 2,400 kilometres (1,500 miles) per hour. Is all that energy coming from the Sun, from the planet's core or from gravitational contraction? Researchers are working to find out.

23 Pluto also has a bizarre atmosphere

Pluto's observed atmosphere broke all the predictions. As data from NASA's New Horizons flowed in, scientists analysed the haze and discovered some surprises. They found about 20 layers in Pluto's atmosphere that are both cooler and more compact than expected. This affects calculations for how quickly Pluto loses its nitrogen-rich atmosphere to space. The New Horizons team found that tonnes of nitrogen gas escapes the dwarf planet by the hour, but somehow Pluto can constantly resupply that lost nitrogen. The dwarf planet is likely creating more of it through geological activity.

24 There are mountains on Pluto

A tiny world at the edge of the Solar System, scientists assumed the dwarf planet would have a fairly uniform, crater-pocked environment. That changed when New Horizons flew by in 2015, sending back pictures that altered our view of Pluto forever. Among the astounding discoveries were icy mountains that are 3,300 metres (11,000 feet) high, indicating that Pluto must have been geologically active as little as 100 million years ago. Geological activity requires energy, and the source of that energy inside Pluto is a mystery. The Sun is too far away from Pluto to generate enough heat for geological activity and there are no large planets nearby that could have caused such disruption with gravity.

25 UNBELIEVABLE FACTS ABOUT THE SOLAR SYSTEM

25 There may be a huge planet at the Solar System's edge

In January 2015, California Institute of Technology astronomers Konstantin Batygin and Mike Brown announced – based on mathematical calculations and simulations – that there could be a giant planet lurking far beyond Neptune. Several teams are now on the search for the theoretical 'Planet Nine', and research suggests it could be located within the decade. This large object, if it exists, could help explain the movements of some objects in the Kuiper Belt, an icy collection of objects beyond Neptune's orbit. Brown has already discovered several large objects in that area that in some cases rivalled or exceeded the size of Pluto, but scientists are pursuing another theory, too – that Planet Nine could in fact be a grapefruit-sized black hole, warping space similarly to the way a gigantic planet would. Yet another team suggests that the weird movements of the far-flung Kuiper Belt occupants could be the collective influence of several small objects, not an undiscovered planet or black hole at all.

"Several teams are now on the search for the theoretical 'Planet Nine'"

The Sun and planets

UNDERSTANDING THE SOLAR SYSTEM

The Sun

The Solar System would be nothing if it weren't for the power and influence of our nearest star

The Milky Way is home to billions of stars, and the universe is home to billions of galaxies. Our star, the Sun, has created a residential spot we call the Solar System. The Sun is the true centre of the Solar System: not only does everything else orbit around it, from asteroids to gas giants, but it makes up 99.8 per cent of the Solar System's mass and is 108 times the diameter of Earth.

This colossus was formed 4.6 billion years ago from a cloud of dust and gas. After this cloud began to rotate and collapse, it took 50 million years to form and become the star it is today. This was the point where the core of the star reached pressures and temperatures so intense that nuclear fusion was ignited. This ignition kick-started the formation of helium from hydrogen, releasing radiation that provides Earth with light, warmth, power and so much more.

At the start of the Sun's life, planets and everything else in orbit were still forming from the leftover debris that began to fragment. The Sun has been in its mature state for nearly 5 billion years, and it will continue to be for another 5 billion years. After this the hydrogen will run out and the Sun will look to form heavier elements such as carbon, oxygen and so on. When this happens the radiation output will be greater than its gravity and it will swell into a red giant, swallowing Mercury and Venus and evaporating all the water and life on Earth. Afterwards the outer layers will be expelled into the cosmos, and what will be left is a white dwarf star.

The Sun has several zones within its interior and atmosphere, starting with a core that burns at over 15 million degrees Celsius (27 million degrees Fahrenheit). This takes up roughly a quarter of the distance to the surface, and outside the core are the radiation and then the convection zones. The coldest layer of the Sun is the photosphere – the visible surface. This is between 6,125 and 4,125 degrees Celsius (11,000 and 7,460 degrees Fahrenheit). Next is the chromosphere and the mysterious corona, invisible without the aid of an eclipse. The corona's temperature ranges from 1 to 10 million degrees Celsius (1.7 to 17 million degrees Fahrenheit), and is perplexing to astronomers because it gets hotter the further away from the Sun you are.

The activity of the Sun creates a magnetic field that permeates the Solar System. Because the Sun is essentially a ball of plasma – matter consisting of ionised gas – and not a solid, it rotates at different speeds depending on its latitude. This unequal rotation causes kinks and twists in the magnetic field, creating sunspots and solar flares, which are usually accompanied by the expulsion of energetic particles in the form of solar wind and coronal mass ejections (CMEs). These particles provide Earth with its aurorae as they collide with the Earth's upper atmosphere.

Observing the Sun in different wavelengths can reveal more about its surface and atmosphere

THE SUN

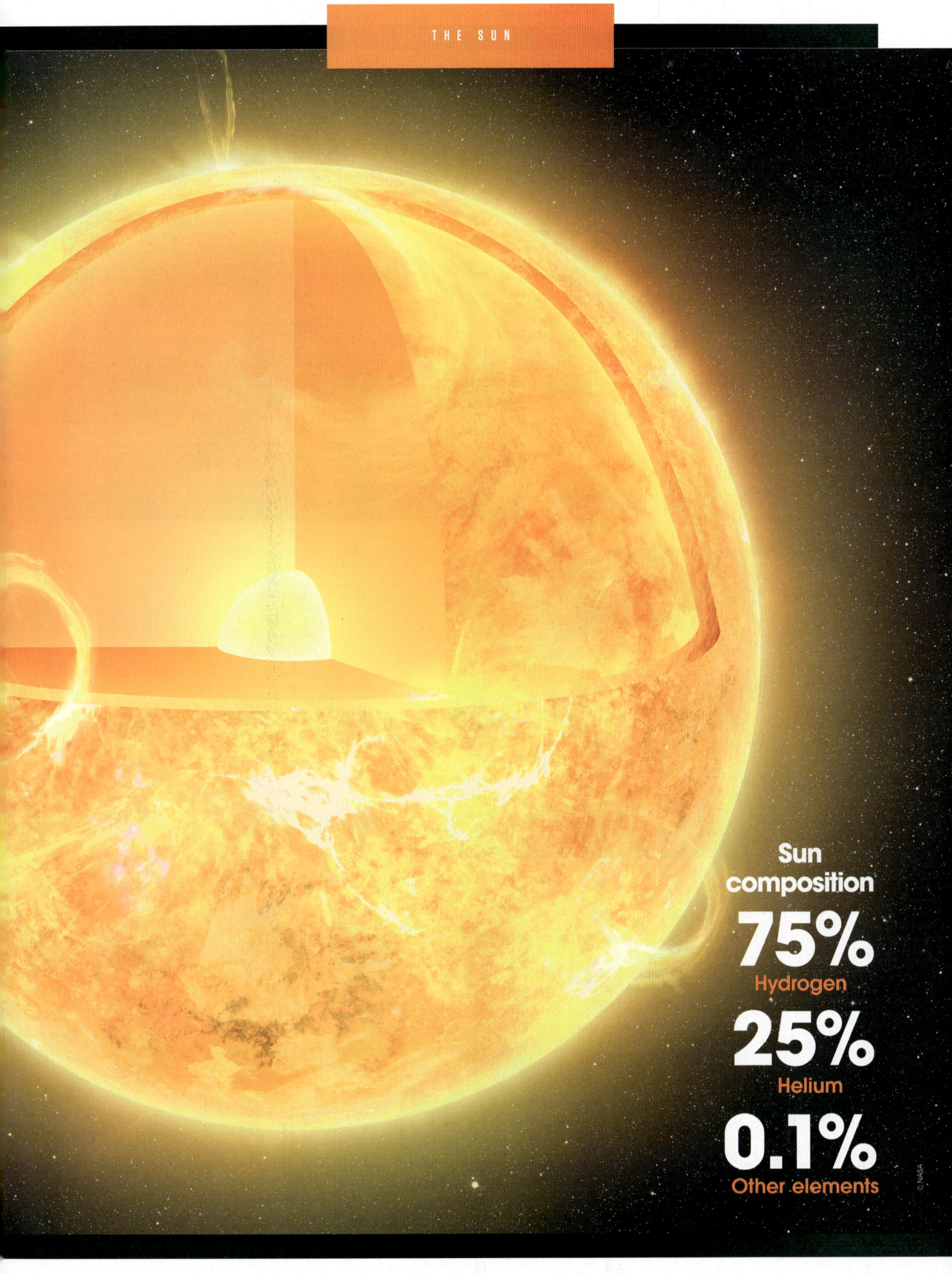

Sun composition

75%
Hydrogen

25%
Helium

0.1%
Other elements

UNDERSTANDING THE SOLAR SYSTEM

News from the Sun
Explore some of the features of our Solar System's star

Studying the solar wind

Scientific papers have been released based on the data collected by NASA's Parker Solar Probe (PSP). One of the new announcements shows the complexity of solar wind. On Earth it is seen as a constant flow of plasma, but the PSP has shown a more complex and active system. The PSP's FIELDS instrument has found sudden reversals in the magnetic field and incredibly fast jets of material occurring much closer to the Sun. Understanding this is key to understanding how the solar wind is moving away from the Sun and permeates the Solar System. "The complexity was mind-blowing when we first started looking at the data," says Stuart Bale of the University of California, Berkeley, lead for the PSP's FIELDS instrument suite.

What's happening to the dust?

Dust is everywhere in space. However, there is a dust-free zone close to the Sun, and it's thought the temperatures up close obliterate it – but this has never been proven. Now the PSP has shown evidence that this is the case. Its Wide-field Imager for Solar Probe (WISPR) has observed dust thinning at a distance of just over 11 million kilometres (7 million miles) from the Sun. The thinning effect continued to be observed as close as 6.4 million kilometres (4 million miles). "This dust-free zone was predicted decades ago, but has never been seen before," says Russ Howard, principal investigator for the WISPR suite at the Naval Research Laboratory in Washington, DC. "We are now seeing what's happening to the dust near the Sun."

Detecting solar flares in real time

Scientists at the Cooperative Institute for Research in Environmental Sciences (CIRES) and the National Oceanic and Atmospheric Administration's (NOAA) National Centers for Environmental Information have developed a machine-learning technique to highlight significant changes in space weather. "Being able to process solar data in real time is important because flares erupting on the Sun impact Earth over the course of minutes. These techniques provide a rapid, continuously updated overview of solar features and can point us to areas requiring more scrutiny," says Rob Steenburgh, a forecaster at the NOAA Space Weather Prediction Center in Boulder, Colorado.

THE SUN

Sun facts

1.3 million Earths could sit within the Sun, or around **1,000 Jupiters**

The Sun contains traces of heavier elements such as **oxygen, carbon, neon** and **iron**

At the equator of the Sun, it takes **25 days** to complete one full rotation, but regions towards the poles take closer to **36 days**

The shape of the Sun relies upon hydrostatic equilibrium, which means the pressure of gravity is equal to radiation output

Sunspots are darker, cooler spots that appear on the Sun's photosphere that arise from complications within the star's magnetic field

An astronomical unit (AU) is the distance between the Sun and Earth and is used to describe distances in the Solar System

Honourable mentions for solar explorers

Mission: Helios
Operator: NASA/German Aerospace Center (DLR)
Active years: 1974 to 1986

Mission: Ulysses
Operator: NASA/ESA
Active years: 1990 to 2009

Mission: Solar and Heliospheric Observatory (SOHO)
Operator: NASA/ESA
Active years: 1995 to present

Mission: Genesis
Operator: NASA
Active years: 2001 to 2004

Mission: Deep Space Climate Observatory (DSCOVR)
Operator: NASA/NOAA
Active years: 2015 to present

Mission: Parker Solar Probe
Operator: NASA
Active years: 2018 to present

Mission: Solar Orbiter
Operator: ESA
Active years: 2020 to present

> "The coldest layer of the Sun is the photosphere – the visible surface"

The past, present and future of solar exploration

There have been many satellites launched to investigate the Sun's activity, the first dating back to 1960 with NASA's Pioneer 5 spacecraft. Since then, instruments, engineering and our understanding of the solar environment have drastically improved. A few missions that have shaped understanding include NASA and the European Space Agency's (ESA) Solar and Heliospheric Observatory (SOHO), NASA's Solar Dynamics Observatory (SDO) and the Japan Aerospace Exploration Agency (JAXA) mission – with collaboration from NASA and the United Kingdom – called Hinode.

On 12 August 2018, NASA launched its Parker Solar Probe (PSP), breaking boundaries when it comes to scrutinising the Sun. At its closest approach the PSP will travel within the Sun's atmosphere at a distance of 3.8 million miles. This mission carries with it four specially designed instrumental suites that look to answer questions about the corona and solar wind while experiencing temperatures of roughly 1,377 degrees Celsius (2,500 degrees Fahrenheit).

In 2020, the ESA launched its Solar Orbiter to get close to the Sun (but not as close as the PSP) at a distance of 41.8 million kilometres (26 million miles). The main difference is that the Solar Orbiter will utilise the gravity of Venus to swing it into a greater inclination, potentially as high as 33 degrees. This will allow it to probe the poles of the Sun, a feat that has never before been accomplished by any other spacecraft.

UNDERSTANDING THE SOLAR SYSTEM

Scientists get a gruesome look at how our Sun will eat planets

Peering into our Solar System's future

One day our Sun will expand into a red giant and engulf its closest planets, and a new study now explores how these devoured planets can influence the processes inside the dying star. When stars the size of our Sun run out of hydrogen in their cores, they balloon into red giants that can be more than ten times larger than the original star. As these red giants engulf the planets that orbit them, many things can happen. Engulfing large planets, ten or more times the size of Jupiter, can trigger the star into shedding its envelope and increasing its brightness by several orders of magnitude for several thousands of years.

The study was conducted using a method called hydrodynamical simulations and provides a glimpse into the possible future scenarios of our own Solar System's evolution. Because of the size of red giant stars, the researchers had to model only a small section of the boundary where the stars meet the planets to gain in-depth insights into the interactions. "Evolved stars can be hundreds or even thousands of times larger than their planets, and this disparity of scales makes it difficult to perform simulations that accurately model the physical processes occurring at each scale," said Ricardo Yarza, a graduate astronomy student at the University of California, Santa Cruz, and lead scientist of the study. "Instead, we simulate a small section of the star centred on the planet to understand the flow around the planet and measure the drag forces acting on it."

Not only could the results provide a glimpse into what will happen 5 billion years from now when our Sun turns into a red giant, but they could also explain recent findings of planets orbiting white dwarfs, the burned-out stellar corpses into which stars turn after the red giant phase. These studies, exploring the end stages of this planetary engulfment, suggest that some planets may survive being burnt by the red giants.

In our Solar System, the closest planets to the Sun – Mercury and Venus – are expected to get swallowed by the growing Sun entirely. Earth, while it may survive, will be so scorched that it will become completely uninhabitable. Some of the more distant and currently freezing cold bodies, such as Jupiter, Saturn and their moons, may develop more life-friendly conditions in the vicinity of the blown-up Sun. While only a few planets that have likely survived a red giant engulfment have been observed so far, researchers believe that further studies of exoplanets will lead to more such discoveries.

Earth will be so scorched by the expanding Sun that it becomes uninhabitable

THE SWELLING SUN

1 Mercury
If Mercury's orbit remains stable until the Sun's red giant phase, the innermost planet is sure to be swallowed up by the expanding star.

2 Venus
The Sun's increasing brightness will blast away Venus' present-day thick atmosphere before swallowing up the planet itself.

3 Earth
The Sun's loss of mass will send Earth's orbit spiralling outwards, but then tidal effects will draw it back, ensuring that our planet, too, is burnt to a cinder in the Sun's outer layers.

4 Mars
Our outer neighbour will probably survive the Sun's expansion and could even have habitable conditions for 100 million years or so.

5 Jupiter
Increased solar heating will cause Jupiter's atmosphere to balloon in size. Its giant moons will enjoy a habitable phase lasting around 30 million years.

6 Saturn
Saturn's atmosphere will also expand, and its rings will likely be long gone. Its moons, including the giant Titan, may benefit from 8 million years at habitable temperatures.

7 Uranus
The inner of the Solar System's two ice giants, Uranus will experience habitable temperatures, warm enough to melt ice on the surface of its moons, for just 2 million years.

8 Neptune
The outermost of the major planets, Neptune orbits so far from the Sun that it will never make it out of deep-freeze, even when the Sun is at its biggest and brightest.

> "In our Solar System, Mercury and Venus are expected to get swallowed by the growing Sun"

UNDERSTANDING THE SOLAR SYSTEM

Mercury

This minute world is arguably the least explored of the four terrestrial planets

Mercury is the smallest of all the planets in the Solar System and the closest planet to the Sun, but there's so much more to it. Mercury is so tiny compared to the other planets that you can actually fit around 23,500 Mercurys into Jupiter, though it's roughly 1,400 kilometres (870 miles) larger in diameter than the Moon.

The small planet orbits the Sun with less than half the distance between the Sun and Earth, resulting in it being 'tidally locked'. Tidal locking occurs when an object is so close to its host that the gravity is overwhelmingly powerful; because of this influence, instead of continuously spinning on its axis like Earth does, the object has one side facing towards its host object at all times. In this case Mercury is tidally locked to the Sun. For every two revolutions around the Sun, Mercury rotates on its axis three times. Each orbit takes 88 Earth days, making a year on Mercury roughly a quarter of an Earth year.

As Mercury is so close to the Sun, the surface temperatures can be scorching, reaching highs of 450 degrees Celsius (840 degrees Fahrenheit). Enduring this bombardment of solar radiation, the planet also struggles to keep hold of its atmosphere, meaning that no heat is trapped. This means the nightside of the planet – the one facing away from the Sun – can have temperatures as low as -180 degrees Celsius (-290 degrees Fahrenheit).

While Mercury is a similar size to the Moon, it's also similar in appearance. It's a heavily cratered, rocky body with some of the largest craters in the Solar System. One such crater studied by previous exploration missions is a great example. The Caloris Basin, which is roughly 1,550 kilometres (960 miles) wide, is about the size of Texas and was formed when an asteroid about 100 kilometres (60 miles) across hit Mercury's surface 4 billion years ago, impacting the planet with energy equivalent to a trillion one-megatonne bombs.

If you scratch beneath the surface, the true weirdness of Mercury starts to become apparent. Under the ultra-thin cratered crust is an extremely dense planet, with somewhere between 70 and 85 per cent of the planet being an enormous iron core. Astronomers have spent years constraining whether it's solid, molten or both, and they seem to agree it has a solid iron core with an outer molten core. Astronomers believe that a molten core explains Mercury's very weak magnetic field. After data was brought back from NASA's Mariner 10 and MESSENGER (Mercury Surface, Space Environment, Geochemistry and Ranging) space probes and analysed, astronomers posited that Mercury is the exposed core of a much larger planet, with its outer layers lost to a powerful collision billions of years ago.

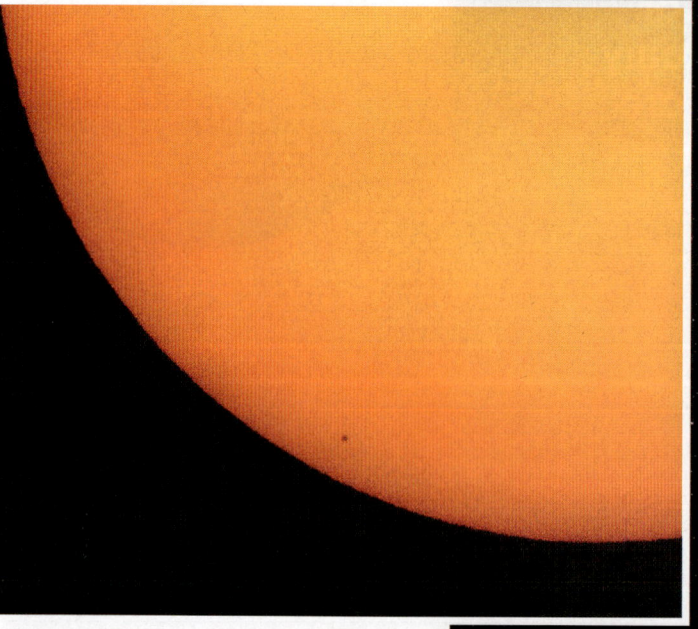

During its 88-day orbit, Mercury will sometimes pass in front of the Sun as viewed from Earth in a planetary transit

MERCURY

The surface of Mercury is pockmarked with craters of various sizes

Atmospheric composition

42%
Oxygen

29%
Sodium

22%
Hydrogen

6%
Helium

0.5%
Potassium

+ Traces of argon, carbon dioxide, water, nitrogen, xenon, krypton, neon, calcium and magnesium

© NASA

25

News from Mercury

Discover some of the fascinating features of the Solar System's smallest planet

Magnetic irregularities

A magnetic field is the result of the motion of a molten core. Earth's magnetic poles have been known to shift, but astronomers have suggested that Mercury's have been doing the same. Mercury's ancient magnetic poles, known as palaeopoles, appear to have shifted over time, and could present clues in the investigation of Mercury's interior. By understanding the magnetic field, we could pinpoint the nature of the planet's molten core. Results came from MESSENGER data on ancient craters that had irregular magnetic signatures. Not only would a further analysis help us understand the nature of Mercury's interior, it could have implications for understanding how the planet evolved, and even how Earth's magnetic field evolved.

Insulating iron sulphide

Once again Mercury's magnetic field is the centre of research. Instead of trying to understand its nature, astronomers are trying to understand how it's kept in place. Astronomers have seen with Mars how a planet smaller than Earth can solidify its molten core, consequently losing its magnetic field, but Mercury still appears to have one. Recent research suggests that a layer of iron sulphide could be insulating the core, maintaining its molten state. Experiments predict that Mercury has a solid inner core with a molten outer core of iron, sulphur and silicon. These elements can't mix, so the iron and sulphur compounds were expelled towards the outer regions of the planet, creating an insulating layer.

Searching for water ice

Mercury may be hiding water at its poles, with MESSENGER revealing the signatures of thick deposits of water ice hidden in craters at the planet's poles. Sunlight doesn't reach the craters' depths, so they're sheltered from the radiation that causes water ice to dissipate. Astronomers believe these craters could hold answers about where water is dispersed in the Solar System. Astronomers have also been comparing Mercury and the Moon to try and understand what the water ice in these craters may look like. This involved looking at around 14,000 craters on the two bodies. The conclusion was that on Mercury, craters that harbour ice have shallower sides than those that don't.

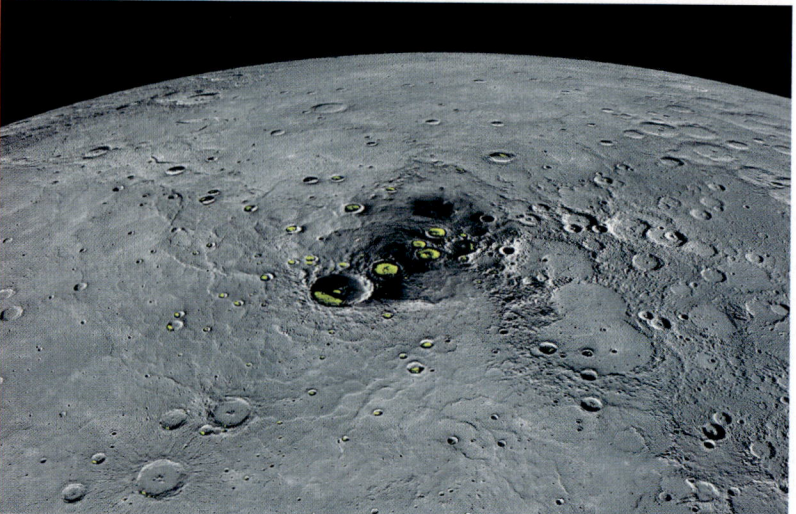

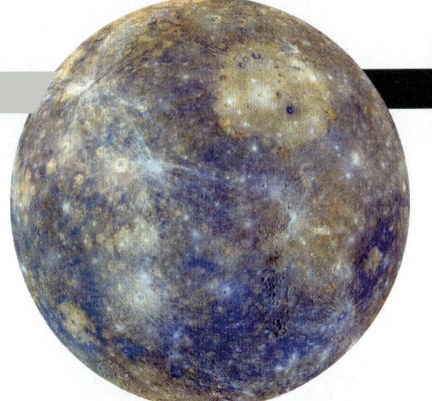

MERCURY

BepiColombo's seven-year journey to Mercury

Date: 20 October 2018
Activity: Launch from Earth.

Date: 10 April 2020
Activity: Earth flyby.

Date: 15 October 2020
Activity: First Venus flyby.

Date: 10 August 2021
Activity: Second Venus flyby.

Date: 1 October 2021
Activity: First Mercury flyby.

Date: 9 January 2025
Activity: Sixth and final Mercury flyby.

Date: 5 December 2025
Activity: Orbital insertion around Mercury.

Mercury facts

NASA's **MESSENGER** mission stayed in orbit around Mercury from March 2011 to April 2015 before crashing into the surface of the planet

On Mercury's nightside at the right time of year, there's a faint orange glow from the sodium scattered by sunlight

59
One day on Mercury lasts 59 Earth days, while a year on Mercury lasts just 88 Earth days

0
Mercury is one of two planets in the Solar System with no moons

Mercury's atmosphere is more comparable to a 'thin exosphere' as it's comprised mostly of atoms ejected from the surface due to the solar wind and meteoroid impacts

There was likely volcanism on Mercury – there are areas that appear to have been flooded with lava

Exploring the past and future of the swift planet

Visiting Mercury is a dangerous and difficult task. Navigating a spacecraft here requires propulsion that will get it to Mercury but also counteract the gravity of the Sun so the craft doesn't go falling into its surface and burn up. This is why only two spacecraft have ever visited the small planet. NASA has been the operator of both of these, the first being Mariner 10 in 1974, which conducted a series of flybys and gathered close-up images.

The mission that brought the most fascinating results is MESSENGER – the first and only spacecraft to orbit the planet. MESSENGER's most important results included how volatile-rich the planet was – volatiles being chemical compounds with low boiling points – which has important implications for the planet's formation. It also found ice deposits at the poles, the weird magnetic field offset and irregular depressions called 'hollows'.

BepiColombo, a joint endeavour by the European Space Agency (ESA) and the Japan Aerospace Exploration Agency (JAXA) will arrive at Mercury in 2025, where it will separate into two orbiters and use its impressive instrumental suite to investigate the planet from all angles. This unique mission will have its two orbiters working simultaneously as scientists get up-close observations of the surface and more distant observations of the magnetic field.

UNDERSTANDING THE SOLAR SYSTEM

22 things you didn't know about Venus

Earth's sister planet is an intriguing and mysterious world, with much more to it than meets the eye

VENUS

1 Venus has a rich history
Studies of Venus can be traced back to the ancient Babylonians in 1600 BCE. They tracked the movement of several planets and stars. The oldest astronomical document on record is a Babylonian diary of Venus appearances over a 21-year period. Venus played a serious part in the mythology of ancient civilisations, including the Maya and Greeks. Its name comes from the Roman goddess of love and beauty.

2 The pressure's on
Walking around on Venus would be an unbearable experience for astronauts for several reasons, but one of them is the extreme pressures on the surface. The atmosphere creates air pressure that's over 90 times the air pressure on Earth, which is similar to the pressure around a kilometre (0.6 miles) deep in the ocean.

1 Atmosphere
96.5 per cent is carbon dioxide, with nitrogen, sulphur dioxide, argon, carbon monoxide, helium and more.

2 Crust
Venus' crust is made of silicate rocks and is estimated to be 50 kilometres (31 miles) thick.

3 Metallic core
Venus' iron core consists of a solid inner and liquid outer core 3,200 kilometres (2,000 miles) in radius.

4 Molten mantle
The heat from the core creates a molten mantle that is 3,000 kilometres (1,200 miles) thick.

UNDERSTANDING THE SOLAR SYSTEM

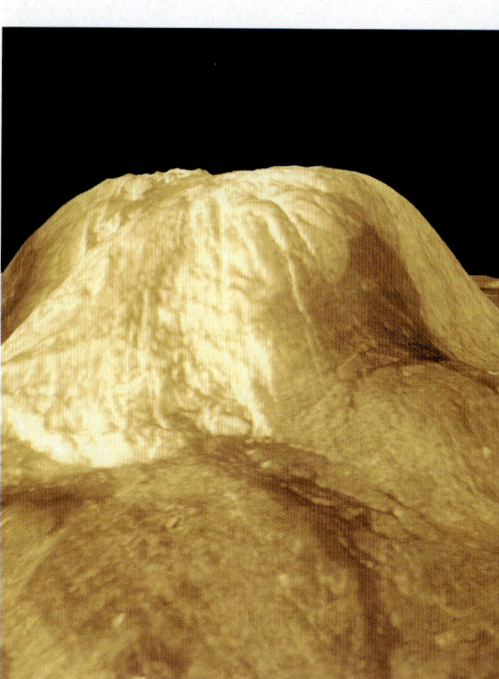

3 It's similar to Earth
When looking purely at the physical parameters of Venus, it's remarkably similar to Earth. They're almost the same in size and density, their compositions are similar and they both appear to have relatively young surfaces that are surrounded by an atmosphere with clouds. It's worth stating that Venus' clouds are primarily sulphuric acid though, which isn't something that you'd want raining down on you.

4 It's just a phase
Venus experiences different phases, just like the Moon. As Venus travels around the Sun within the orbit of Earth, it changes between a 'morning star' and 'evening star' roughly every nine-and-a-half months. During this period it shifts between different percentages of illumination, a trait that is normally associated with the Moon.

5 Transits are very rare
Venus is one of two planets that orbit the Sun within the orbital path of Earth. Along with Mercury, these two planets can find themselves between Earth and the Sun, sometimes creating a silhouette that moves across the Sun over a period of hours. These journeys are known as 'transits', and Venus is known to transit in pairs – though with over a century separating the pairs, it's a very rare event.

6 It's hellishly hot
Venus is the hottest planet in the Solar System, even hotter than the dayside of Mercury, which has temperatures of 427 degrees Celsius (801 degrees Fahrenheit). Because of Venus' thick, carbon dioxide-rich atmosphere, the heat is efficiently retained, creating surface temperatures higher than 470 degrees Celsius (880 degrees Fahrenheit).

7 Venusian volcanicity
To add to the hellish image of Venus, it also has the most volcanoes present on the surface of all the planets in the Solar System. On Earth there are 1,500 known active volcanoes, and Mars is best known for the largest volcano in the Solar System, Olympus Mons. However, Venus has over 1,600 major volcanoes, and that's not including the smaller ones or any that haven't been detected yet.

8 It doesn't have a moon
Venus and Mercury are the only planets in our Solar System that don't have their own moon. It's a bit more understandable why Mercury doesn't have a moon, because its close proximity to the Sun has a negative effect on any contenders. The planet is also smaller than some Solar System moons, such as Saturn's Titan. However, researchers have argued that the reason Venus doesn't have a moon isn't as simplistic. There are two theories: the first is that any moon that Venus had was stolen by the Sun's gravity. The second is known as the 'double-impact theory', which states that a large celestial body hit Venus billions of years ago, creating a moon in a similar way to how Earth got its lunar companion. But several million years later, an even bigger object hit Venus, causing its retrograde rotation, weakening the tidal forces and sending the moon to sink into Venus, never to be seen again.

9 Earth vs Venus

The Sun
On Venus, the Sun would appear no more than a dimly glowing patch through the thick clouds.

Clouds
Venus is enveloped in clouds, not allowing any nosey astronomers to investigate the surface. While Earth is also hidden by clouds, much more of our planet's surface is visible from space.

Surface rocks
Based on past exploration missions, the surface of Venus contains rocks of different shades of grey, carving out valleys and giving birth to mountains, similar to Earth.

Volcanoes
Both planets feature at least 1,500 active volcanoes on the surface, and many more dormant ones.

VENUS

10 A perfect world for futuristic spacecraft
There are advantages to scrutinising Venus from its clouds

Faster exploration
Super-rotation in the upper atmosphere, which completes a rotation 60 times quicker than the surface below, would allow for rapid exploration of Venus.

Improved capabilities
With lightweight technologies and controlled aerial mobility, aircraft on Venus is now a better proposal than it was in the 1960s.

Head in the clouds
There's discussion about whether it would be possible to create a colony in the clouds of Venus, much like Cloud City on Bespin in the Star Wars universe.

Easier to explore from up high
There are more favourable conditions in the clouds, with much more bearable temperatures and pressures.

The power of the Sun
Solar panels would be extremely useful, as Venus gets 190 per cent more sunlight than Earth.

Removing obstacles
Being in constant flight eliminates the need to navigate around harmful terrain and the planet's volcanoes.

"Conditions on Venus that would be favourable for life could exist in the clouds"

11 Life in the clouds
Researchers have proposed that life could be found on Venus, just not on the surface. A study by Sanjay Limaye of the University of Wisconsin-Madison's Space Science and Engineering Center suggested that microbial life could be present in the cloud tops. Microbial life on Earth has been found at altitudes of 41 kilometres (25 miles), and researchers have said that conditions on Venus that would be favourable for life could exist in the clouds at altitudes of 48 to 51 kilometres (30 to 32 miles). Here, temperatures would be roughly 60 degrees Celsius (140 degrees Fahrenheit) and pressures would be similar to Earth at sea level.

12 A day feels like a year
On Venus that's very much the case. One Venusian day, which is one complete rotation on its axis, takes 243 Earth days, making it the longest day of any planet in the Solar System. Even a year on Venus is shorter, as it takes 224.7 Earth days to complete one revolution around the Sun.

13 'Backwards' rotation
Another trait that makes Venus different to most of the planets in the Solar System is its rotation. The usual routine for planets is to spin anticlockwise on their axis, but Venus is an oddball and flaunts a clockwise rotation. The leading theory as to why Venus and Uranus have what is known as a 'retrograde rotation' is that they were smacked by large objects early in their history. This collision left the planet seeing stars and spinning the wrong way.

UNDERSTANDING THE SOLAR SYSTEM

14 What the future holds

Researchers want to understand every planet in the Solar System. Efforts in the late-20th century showed that Venus is a difficult planet to observe remotely from the surface, but with new technologies and a better understanding comes innovative exploration ideas. A lot of these new ideas have a common theme, which is exploring Venus from within the clouds. As Venus has more favourable conditions in the clouds, with wind speeds that allow an object to travel around the planet much faster than it rotates, scientists are looking to introduce aircraft or airships. By utilising solar and wind power, and with the added help of buoyancy, robotic missions could become a feature of Venus in the foreseeable future.

15 Rewinding the clock

Much like Mars, Venus could have once supported life. 700 million years ago, Venus suffered dramatic changes in its climate that saw it bulk up its atmosphere in a process known as a 'runaway greenhouse effect'. Before the runaway greenhouse effect took over, it's believed that Venus had a reasonable atmosphere and could have harboured liquid water for about 2 or 3 billion years. Before carbon dioxide dominated the atmosphere and made it too hot and dense, it's possible that Venus had an environment that could have supported life for billions of years.

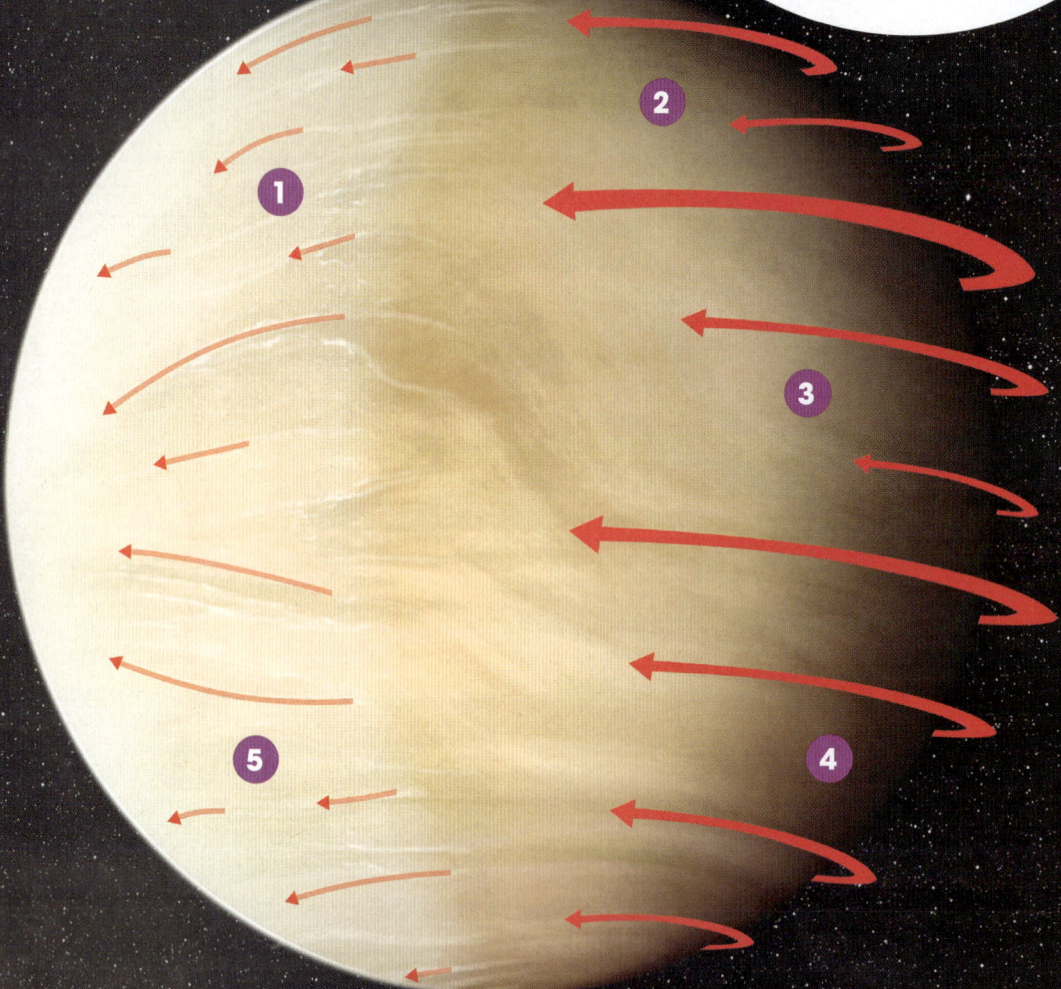

1 Seen from above
These irregular, patchy, filament-like structures were observed by the European Space Agency's Venus Express spacecraft.

2 Stationary waves
Stationary or gravity waves in the nightside's atmosphere do not move in the same way as the planet's super-rotation.

3 Indirect surface observations
These waves come from steep, mountainous areas on Venus that send waves through the atmosphere.

4 Never-ending heat
The extremely slow rotation and tilt of just 3.39 degrees ensure that the planet stays continuously hot.

5 The mystery of the nightside
On the nightside the upper clouds form in different shapes and morphologies, causing a more irregular system.

VENUS

16 Too slow to be magnetic
Although it's often referred to as Earth's twin, something that differentiates the two planets deep down to their cores is that Venus creates a negligible magnetic field. Planetary scientists believe that Venus has an iron core that's a similar size to Earth's. However, the sluggish rotation of Venus, which consequently reduces the motion of the planet's core, weakens the planet's magnetic field, or magnetosphere.

17 It's had many spacecraft visitors
Before attention turned to the exploration of Mars, Venus was where space agencies wanted to send their robotic missions to. This genesis of interplanetary exploration began with a lot of spacecraft and launch failures, starting with the Soviet Union's Tyazhely Sputnik in February 1961, which experienced a launch failure. There have since been 45 other missions launched with the intention of exploring the planet. Of these missions, more than 20 have been successful. The very first to conduct a successful planetary encounter was NASA's Mariner 2 space probe on 14 December 1962.

18 The case of the missing lightning
There are electrical pulses bursting through the heavy atmosphere, but missions to Venus to find them have made things even more confusing. Ground-based telescopes and space probes alike, including the ESA's Venus Express and the Japan Aerospace Exploration Agency's (JAXA) Akatsuki, have had nothing more than some subtle hints about the presence of Venusian lightning. Researchers believe it could still be present, just much more localised and rare, which is why there has been no definitive evidence yet. Or it could be the case that there isn't lightning at all.

19 Soviet success at Venus
Before the dissolution of the Soviet Union in 1991, the country was prominent in Venus exploration missions in the 1970s and 1980s. One historic mission that the Soviets conducted was Venera 7 in December 1970, which became the first mission to land on a different planet. Then, in March 1982, the Venera 13 lander managed to survive Venus' extreme temperatures and pressures for an astonishing two hours.

20 Brightest in the sky
Because Venus is in such close proximity to Earth, it's the third-brightest celestial object in the night sky behind the Sun and Moon. The Latin nickname for Venus, now largely unused, is 'Lucifer', which translates to 'light bringer'. Lucifer is also a name for the devil, which is quite a coincidence considering the hellish conditions on the surface of Venus.

21 A source of shadows
As the third-brightest object in the night sky, it's bright enough to cast shadows on Earth. Only two other celestial objects are capable of this: the Sun and Moon. Very good eyesight is needed to see these Venusian shadows.

22 Weird winds
The clouds move across the atmosphere once every four Earth days, known as super-rotation. This generates speeds of 360 kilometres (224 miles) per hour, surpassing those of the most dangerous hurricanes on Earth. Speeds decrease with cloud height, creating winds that are just a few miles per hour on the surface.

© Tobias Roetsch; Shutterstock; ESA, NASA/JPL; Adrian Mann

UNDERSTANDING THE SOLAR SYSTEM

What are planets like on the inside?

Even among the worlds of our Solar System we see a huge variety of planets

The terrestrial planets – Mercury, Venus, Earth and Mars – are separated into similar layers: a hot dense core, a warm pliable mantle and a cooled rocky crust. Mercury is around 70 per cent metallic and 30 per cent rocky. Its core is thought to comprise as much as 85 per cent of the planet, a liquid heart of iron 4,000 kilometres (2,485 miles) in diameter. This is covered by 600 kilometres (373 miles) of silicon-rich mantle and between 100 and 200 kilometres (62 and 124 miles) of rocky crust. Venus was expected to have a similar structure to Earth, but Venus has little magnetic field. Earth's field is created by rotation and convection in our molten core; Venus does seem to have a molten core, but it doesn't circulate in the same way. One possibility is that Venus' crust doesn't get recycled like Earth's, so the whole core may be of uniform temperature as heat is not escaping to the surface.

Fortunately for us, Earth has a significant magnetic field that protects us from solar radiation, but our core isn't just liquid. The pressure at the centre of Earth is sufficient enough that the iron collected there becomes solid, despite the temperature being around 6,000 degrees Celsius (10,832 degrees Fahrenheit). This has been determined by studying the way seismic waves travel through Earth. The solid inner core is around 1,220 kilometres (760 miles) in diameter. This is surrounded by a liquid layer that is 2,200 kilometres (1,367 miles) deep, which is then topped by 2,900 kilometres (1,802 miles) of mantle and an average crust of 35 kilometres (22 miles).

Mars is also differentiated into layers, with a liquid core, a mantle and a rocky crust. Like Venus, there must not be convection in the core, as Mars has no magnetic field.

Jupiter is the first of the gaseous planets. When it comes to gas giants, there's no sharp dividing line between the planet and its atmosphere. It's thought that Jupiter has a dense core, possibly rocky, surrounded by 'metallic' hydrogen. This is a strange condition predicted to occur under huge pressures, where hydrogen behaves like a dense electrically conductive substance. This layer is thought to cover 78 per cent of the thickness of the planet. It's thought that above this, normal liquid hydrogen smoothly fades into the gaseous hydrogen atmosphere. Saturn is thought to be similar to Jupiter, with a rocky or icy core surrounded by the same types of hydrogen layers.

Uranus and Neptune are called the ice giants, due to the layer of mixed methane, water and ammonia that surrounds their rocky cores – equivalent to the mantle in terrestrial planets. These are known as ices, though they form a hot, dense liquid near the core and fade out into the atmosphere without a defined surface.

Expert: Robin Hague
Robin is a science writer, focusing on space and physics. He is head of launch at Skyrora, coordinating launch opportunities for Skyrora's vehicles.

UNDER THE SURFACE
- ROCK
- MOLTEN ROCK
- ICE
- MOLTEN IRON
- IRON
- LIQUID METALLIC HYDROGEN
- HYDROGEN GAS
- ATMOSPHERE

MERCURY
CRUST
MANTLE
CORE

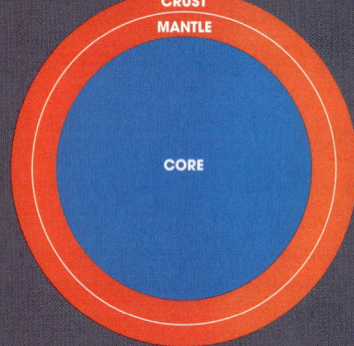

JUPITER
OUTER LAYER
METALLIC HYDROGEN LAYER
CORE*

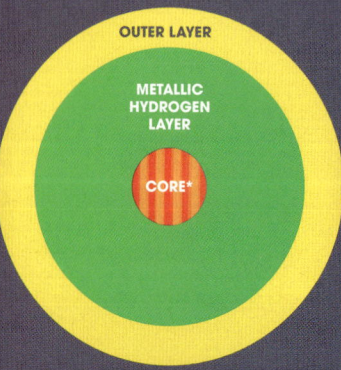

*JUPITER'S CORE REMAINS A MYSTERY TO SCIENTISTS, BUT IT'S HOPED THAT THE JUNO MISSION WILL SHED LIGHT ON ITS SIZE AND COMPOSITION

WHAT ARE PLANETS LIKE ON THE INSIDE?

An artist's impression of water under the Martian surface. If underground aquifers like this exist, the many Mars missions have a good chance of finding them

VENUS
- CRUST
- MANTLE
- CORE

EARTH
- CRUST
- MANTLE
- OUTER CORE
- INNER CORE

MARS
- CRUST
- MANTLE
- CORE

SATURN
- OUTER LAYER
- HYDROGEN GAS LAYER**
- METALLIC HYDROGEN LAYER
- CORE

**THICKNESS UNKNOWN

URANUS
- GAS LAYER
- MANTLE
- CORE

NEPTUNE
- GAS LAYER
- MANTLE
- CORE

UNDERSTANDING THE SOLAR SYSTEM

Earth

The rocky world that we call home is full of wonders

A rather pretty blue-and-white planet orbiting an otherwise obscure G-type main sequence star, Earth is notable largely for being the only place in the universe to have evolved organic life. Other than this quirk of chemistry, the third planet from the Sun also has active plate tectonics, and it's one of the few planets whose moon fits perfectly over its Sun during an eclipse. It is the densest planet in its Solar System, and the largest of the four rocky planets closest to its star. An atmosphere 100 kilometres (62 miles) thick coats the planet, offering it protection from ultraviolet light given out by its nearest star thanks to its layer of ozone. Heating of the upper atmosphere means it's slowly losing its hydrogen and helium into space, but at a very slow rate.

With its thick atmosphere and yellow sunlight, much of Earth's vegetation is green. Its position at around 150 million kilometres (93 million miles) from its star means liquid water is commonplace on its surface – both salty and non-salty forms, freezing at the poles – though a recent increase in atmospheric carbon dioxide levels is causing this ice to melt. Unlike its neighbour Mars, biological life flourishes both in Earth's oceans and on the third of the planet not covered with water.

An axial tilt of 23.5 degrees leads to seasons on Earth, which combine with both atmospheric and oceanic circulations to produce a variety of weather types, some of them extreme. A single natural satellite is tida'ly locked to the planet, and its gravitational pull affects the water level beneath it, causing tides. Along with many artificial satellites, Earth also has a small number of quasi-satellites, mostly captured asteroids circulating around Lagrange points L4 and L5 in horseshoe orbits.

Earth is currently 20,000 years into an interglacial period, part of a cycle of ice ages that sees glaciers coat large parts of the planet over periods of up to 500,000 years. The current interglacial should end in around 25,000 years, though warming caused by increased atmospheric carbon dioxide levels could delay this by trapping heat within the atmosphere.

In a billion years' time, the energy received by Earth from its star will have increased by ten per cent, enough for the oceans to be lost thanks to a combination of subduction into the planet's mantle and photodissociation of the water molecules by increased levels of ultraviolet light. Without surface water, plate tectonics will come to a halt. Earth will become similar to its near-twin Venus, with a runaway greenhouse effect eventually raising the surface temperature to 1,330 degrees Celsius (2,426 degrees Fahrenheit).

In another 5 billion years, the Sun will run out of hydrogen to burn in its core and will begin the process of swelling into a red giant. As it expands, Earth, along with Venus and Mercury, will be engulfed by its chromosphere. Tidal forces will break up the Moon, briefly turning it into a ring system before the surface and mantle are stripped from the Earth, leaving only its core. The final legacy of Earth will be an increase in the Sun's metal content of 0.01 per cent.

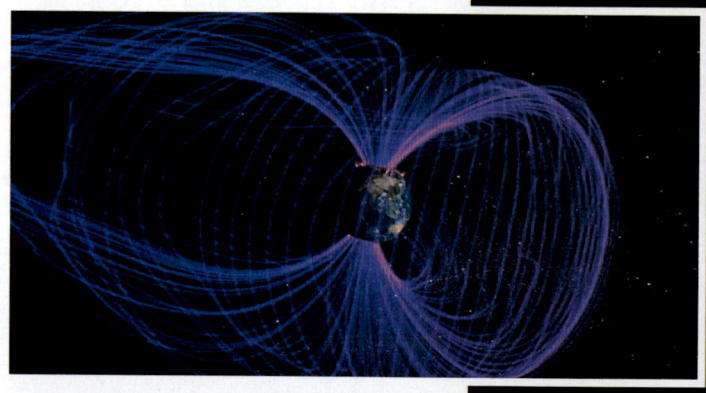

Earth is surrounded by a magnetic bubble called its magnetosphere

EARTH

Atmospheric composition

32.1%
Iron

30.1%
Oxygen

15.1%
Silicon

13.9%
Magnesium

2.9%
Sulphur

1.8%
Nickel

1.5%
Calcium

1.4%
Aluminium

1.2%
Traces of other elements

UNDERSTANDING THE SOLAR SYSTEM

News from Earth
What's happening on our home planet?

Extinction event

A study published in the journal Biological Conservation suggests that 84 per cent of animals and plants in mountain regions risk being wiped out if the temperature rises by more than an average of three degrees Celsius (5.4 degrees Fahrenheit), with this rising to 100 per cent on islands.

Geographically unique species such as Madagascar's lemurs and the snow leopards of the Himalayas are 2.7 times more likely to go extinct than more widespread species. More than 60 per cent of unique tropical species are likely to go extinct thanks to the action of climate change, and places such as the Caribbean islands and Sri Lanka could lose most of their endemic plants by 2050. Up to 92 per cent of species on land and 95 per cent of those in the sea could face negative consequences.

The researchers – from Brazil, Norway and South Africa, among others – concluded that if the world can keep global average temperature rises within the terms of the Paris Climate Agreement, then the risk to vulnerable species drops by a factor of ten. With a 1.5 degrees Celsius (2.7 degrees Fahrenheit) rise, only two per cent of land and marine species face extinction.

Artificial island

Not content with all the islands already available to it, the dominant mammal species on Earth has been busy making more. A new artificial island near Malé, the capital of the Republic of Maldives, an archipelago in the Indian Ocean, will act as a refuge for people stranded by rising sea levels. With more than 80 per cent of its 1,190 islands just one metre (3.2 feet) above the water, the Maldives has the lowest terrain of any country in the world, which makes it particularly susceptible to sea-level rise. Construction of the new island, known as Hulhumalé, began in 1997, and it has grown to over four square kilometres (1.5 square miles) in area. It sits two metres (6.5 feet) above sea level, constructed from sand pumped on top of submerged coral, and is now the fourth-largest island in the archipelago.

With sea levels predicted to rise by up to half a metre by 2100 even if the Paris Climate Agreement targets are hit, land reclamation projects like this may become more common as populations are driven from low-lying areas.

Moon telescope

An early stage proposal has received funding from NASA to build a radio telescope in a crater on the far side of the Moon. Similar in concept to the Arecibo Observatory, the Lunar Crater Radio Telescope would take advantage of the Moon's many meteor craters to support its structure.

Because of the way Earth and the Moon are tidally locked, one side of the Moon always faces away from us. The advantage of building such a device on the Moon, particularly on its far side, is the shielding effect it gives against Earth-generated noise and even the radio waves emitted by the Sun. It would also be able to observe the universe at frequencies that are blocked by Earth's atmosphere, such as those below 30MHz. Observations in these wavebands have never been made by humans.

The proposal is to deploy two wall-climbing robots in a crater three to five kilometres (1.8 to 3.1 miles) in diameter. The robots would then weave a dish one kilometre (0.62 miles) across using a wire mesh. A receiver would then be suspended above this dish on two crossed cables, each end held by a movable robot, adjusting the position of the receiver for the best results.

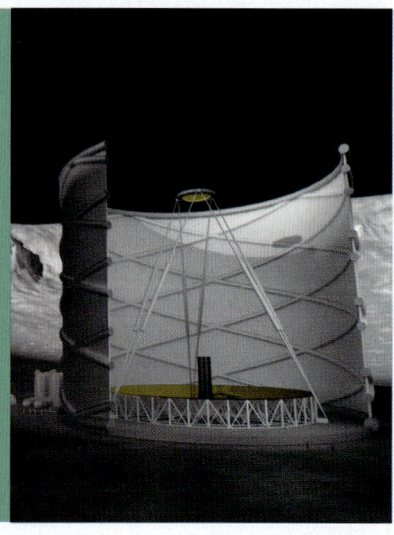

EARTH

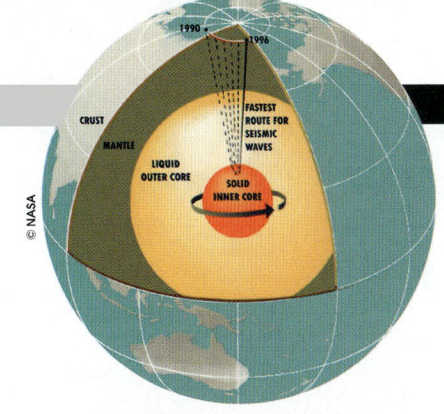

The Evolution of Planet earth

Date: 54 billion years ago
Activity: Earth formed from a protoplanetary disc around a young star.

Date: 4.5 billion years ago
Activity: Dense elements sank to the centre, forming Earth's core, while the outside layer cooled and solidified.

Date: 4.48 billion years ago
Activity: A massive impact with another body sent a portion of Earth's crust into orbit, forming the Moon.

Date: 4.4 billion years ago
Activity: Volcanism released water vapour into Earth's atmosphere, raining down to begin the formation of oceans.

Date: 3.5 billion years ago
Activity: Earth's magnetic field was established, with a magnetosphere about half the modern radius.

Date: 750 million years ago
Activity: The earliest known supercontinent, Rodinia, began to break apart.

Date: 180 million years ago
Activity: The most recent supercontinent, Pangaea, broke apart.

Date: 65 million years ago
Activity: Formation of the Himalayas began as the Indian subcontinent drifted into Asia.

Date: 6 million years ago
Activity: A small African ape began a family tree that led to a dominant species.

Earth facts

43 KM The difference in the Earth's diameter at the equator than if measured pole-to-pole

One natural satellite

14°C Average surface temperature

1 AU Average distance to Sun

5,430°C Temperature at inner core

1 G Average surface gravity

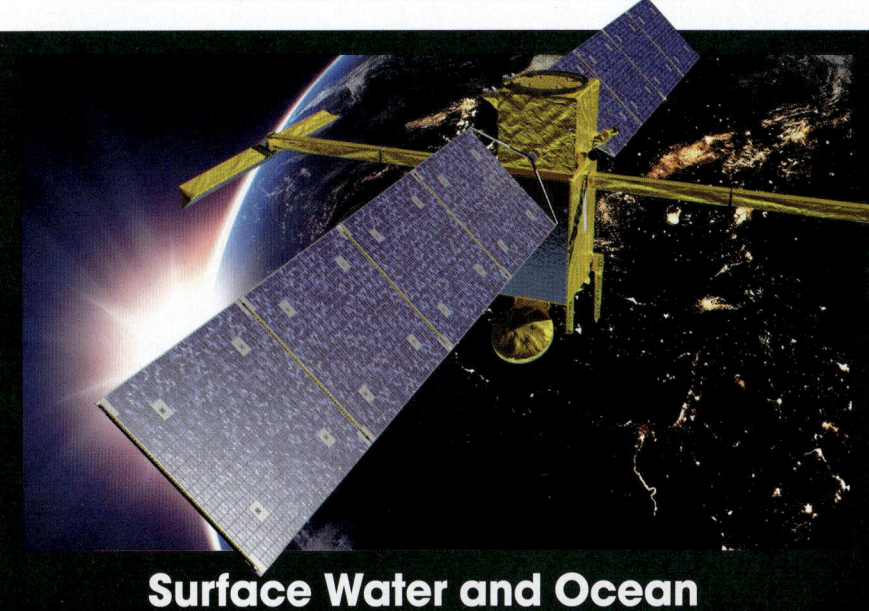

Surface Water and Ocean Topography mission

Launched in December 2022, SWOT – a joint development between NASA and France's CNES agency, with help from Canada and the UK – is set to accurately measure the height of Earth's surface water. The mission aims to measure how bodies of water change over time. It will use a radar altimeter to measure the height of oceans, rivers and lakes across 90 per cent of the globe at least twice every 21 days at an average precision better than 1.5 centimetres (0.6 inches).

This data will lead to better weather and climate forecasting, providing more accurate information about sea and river levels that can be plugged into the supercomputer prediction models used by meteorological agencies. It will also be able to measure the 3D shape of floodwater, track flood levels and improve our ability to predict floods.

The largest effect SWOT may have on Earth's population is the data it will provide about freshwater management. This will help urban planners to manage the distribution of water for agricultural, industrial and urban needs by providing information about reservoirs and rivers. The knowledge we will gain of Earth's water cycle and ocean circulations will help us to better understand everything from surface water to the deep oceans, and this should improve our reactions to natural disasters, waterborne diseases, sharing water sources among different populations, as well as managing electricity production from renewable means and safeguarding biodiversity.

UNDERSTANDING THE SOLAR SYSTEM

Complete guide to Mars

We're learning more about the fourth rock from the Sun every day

MARS

Mars is the fourth planet from the Sun. Befitting the Red Planet's bloody colour, the Romans named it after their god of war. The bright rust colour Mars is known for is due to iron-rich minerals in its regolith – the loose dust and rock covering its surface. The soil of Earth is a kind of regolith, too, albeit one loaded with organic content. The iron minerals on Mars oxidise, or rust, causing its soil to look red.

The planet's thin atmosphere means liquid water likely cannot exist on the Martian surface for any appreciable length of time. Features called recurring slope lineae may have spurts of briny water flowing onto the surface, but this evidence is disputed; some scientists argue the hydrogen spotted from orbit in this region may instead indicate briny salts. This means that although this desert planet is just half the diameter of Earth, it has the same amount of dry land.

The Red Planet is home to both the highest mountain and the deepest, longest valley in the Solar System. Olympus Mons is roughly 25 kilometres (16 miles) high, about three times as tall as Mount Everest, while the Valles Marineris system of valleys – named after the Mariner 9 probe that discovered it – reaches as deep as ten kilometres (six miles) and runs east to west for roughly 4,000 kilometres (2,500 miles), about one-fifth the distance around Mars and close to the width of Australia.

Scientists think that Valles Marineris formed mostly by rifting of the crust as it got stretched. Individual canyons within the system are up to 100 kilometres (60 miles) wide. The canyons merge in the central part of Valles Marineris in a region 600 kilometres (370 miles) wide. Large channels emerging from the ends of some canyons and layered sediments within suggest that the canyons might once have been filled with liquid water. Mars also has the largest volcanoes in the Solar System – Olympus Mons being one of them. The massive volcano, which is about 600 kilometres (370 miles) in diameter, is wide enough to cover the state of New Mexico. Olympus Mons is a shield volcano, with slopes that rise gradually like those of Hawaiian volcanoes, and was created by eruptions of lava that flowed for long distances before solidifying. Mars also has many other kinds of volcanic landforms, from small, steep-sided cones to enormous plains coated in hardened lava. Some minor eruptions might still occur on the planet today.

Channels, valleys and gullies are found all over Mars, suggesting that liquid water might have once flowed across the planet's surface. Some channels are 100 kilometres (60 miles) wide and 2,000 kilometres (1,200 miles) long. Water may still lie in cracks and pores in underground rock. A 2018 study suggested that salty water below the Martian surface could hold a considerable amount of oxygen, which could support microbial life. But the amount of oxygen depends on temperature and pressure; temperature changes on Mars from time to time as the tilt of its rotation axis shifts.

Many regions of Mars are flat, low-lying plains. The low northern plains are among the flattest, smoothest places in the Solar System, potentially created by water that once flowed across the Martian surface. The northern hemisphere mostly lies at a lower elevation than the southern hemisphere, suggesting the crust may be thinner in the north than in the south. This difference between the north and south might be due to a very large impact shortly after the birth of Mars.

The number of craters on Mars varies from place to place, depending on how

Mars statistics

Diameter: 6,791 kilometres (4,220 miles)
Gravity: 38 per cent of Earth's
Atmospheric composition: 95.32% carbon dioxide; 2.7% nitrogen; 1.6% argon; 0.13% oxygen; 0.08% carbon monoxide; traces of water, nitrogen oxide, neon, heavy water, krypton and xenon
Chemical composition:
Core: Iron, nickel and sulphur
Mantle: Peridotite (silicon, oxygen, iron and magnesium)
Crust: Volcanic rock basalt
Moons: Phobos and Deimos

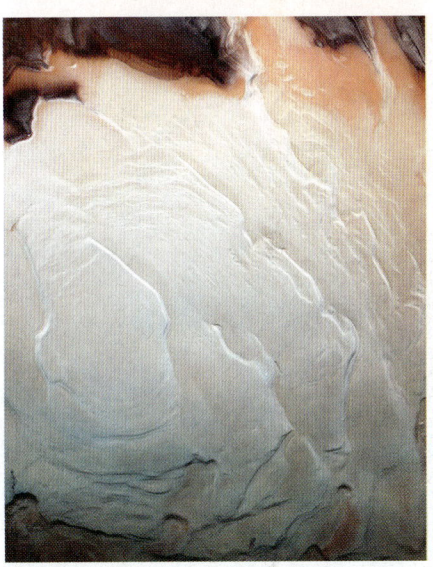

The ice cap which covers Mars' southern pole, composed of water and carbon dioxide

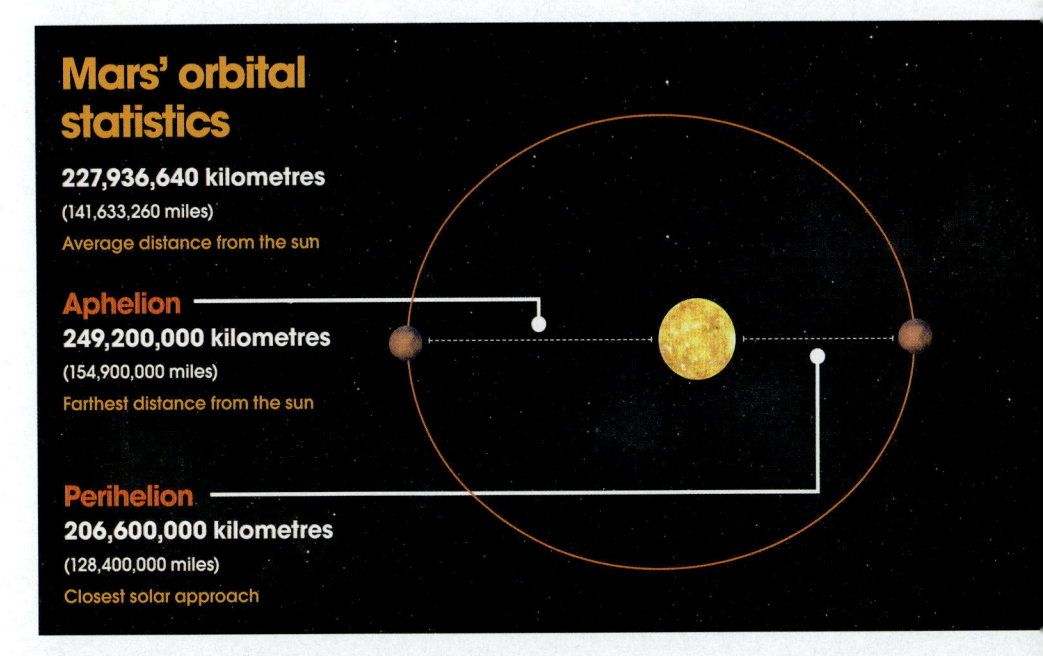

Mars' orbital statistics

227,936,640 kilometres
(141,633,260 miles)
Average distance from the sun

Aphelion
249,200,000 kilometres
(154,900,000 miles)
Farthest distance from the sun

Perihelion
206,600,000 kilometres
(128,400,000 miles)
Closest solar approach

UNDERSTANDING THE SOLAR SYSTEM

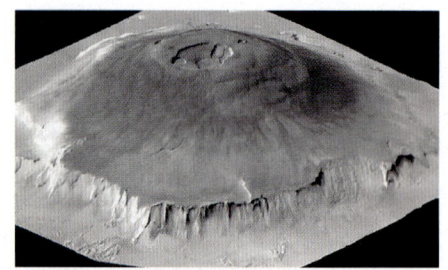

A 3D image of Olympus Mons, the largest known mountain in the Solar System

old the surface is. Much of the surface of the southern hemisphere is extremely old, and so has many craters – including the planet's largest, 2,300-kilometre (1,400-mile) Hellas Planitia – while the northern hemisphere is younger and so has fewer craters. Some volcanoes also have just a few craters, which suggests they erupted recently, with the resulting lava covering up any old craters. Some craters have unusual deposits of debris around them resembling solidified mudflows, indicating that the impactor may have hit underground water or ice.

In 2018, the European Space Agency's (ESA) Mars Express spacecraft detected what could be a slurry of water and grains underneath the icy Planum Australe. This body of water is said to be about 20 kilometres (12.4 miles) across. Its underground location is reminiscent of similar underground lakes in Antarctica, which have been found to host microbes. Mars Express also spied a huge, icy zone in Korolev crater. Vast deposits of what appear to be finely layered stacks of water ice and dust extend from the poles to latitudes of about 80 degrees in both Martian hemispheres. These were probably deposited by the atmosphere over long spans of time. On top of much of these layered deposits in both hemispheres are caps of water ice that remain frozen year-round.

Additional seasonal caps of frost appear in the wintertime. These are made of solid carbon dioxide, also known as 'dry ice', which has condensed from carbon dioxide gas in the atmosphere – Mars' thin air is about 95 per cent carbon dioxide by volume. In the deepest part of the winter, this frost can extend from the poles to latitudes as low as 45 degrees, or halfway to the equator. The dry ice layer appears

Key Mars missions
Robots have been unlocking Mars' secrets for decades

Seismic activity

InSight lander
Launch date: 5 May 2018
Arrival: 26 November 2018

Relay network

Mars Odyssey
Launch date: 7 April 2001
Arrival: 24 October 2001

Mars Express
Launch date: 2 June 2003
Arrival: 25 December 2003

Mars Reconnaissance Orbiter (MRO)
Launch date: 12 August 2005
Arrival: 10 March 2006

Mars Atmospheric and Volatile EvolutioN (MAVEN)
Launch date: 18 November 2013
Arrival: 22 September 2014

Trace Gas Orbiter (TGO)
Launch date: 14 March 2016
Arrival: 19 October 2016

Mars rovers

Spirit
Launch date: 10 June 2003
Arrival: 4 January 2004

Opportunity
Launch date: 7 July 2003
Arrival: 25 January 2004

Curiosity
Launch date: 26 November 2011
Arrival: 6 August 2012

Perseverance
Launch date: 30 July 2020
Arrival: 18 February 2021

to have a fluffy texture, like freshly fallen snow. Mars is much colder than Earth, in large part due to its greater distance from the Sun. The average temperature is about -60 degrees Celsius (-80 degrees Fahrenheit), although it can vary from -125 degrees Celsius (-195 degrees Fahrenheit) near the poles during the winter to as much as 20 degrees Celsius (70 degrees Fahrenheit) at midday near the equator.

The carbon-dioxide-rich atmosphere of Mars is about 100 times less dense than Earth's on average, but it is nevertheless thick enough to support weather, clouds and winds. The density of the atmosphere varies seasonally, as winter forces carbon dioxide to freeze out of the Martian air. In the ancient past, the atmosphere was likely significantly thicker and able to support water flowing on the planet's surface. Over time, lighter molecules in the Martian atmosphere escaped under pressure from the solar wind, which affected the atmosphere because Mars doesn't have a global magnetic field. This process is being studied today by NASA's Mars Atmosphere and Volatile Evolution (MAVEN) mission. NASA's Mars Reconnaissance Orbiter found the first definitive detections of carbon-dioxide snow clouds, making Mars the only body in the Solar System known to host such unusual winter weather. The Red Planet also causes water-ice snow to fall from the clouds.

The dust storms on Mars are the largest in the Solar System, capable of blanketing the entire planet and lasting for months. One theory as to why dust storms can grow so big on Mars is because the airborne dust particles absorb sunlight, warming the Martian atmosphere in their vicinity. Warm pockets of air then flow towards colder regions, generating winds. Strong winds lift more dust off the ground, which in turn heats the atmosphere, raising more wind and kicking up more dust. These storms can pose serious risks to robots on the Martian surface. NASA's Opportunity rover 'died' after being engulfed in a giant 2018 storm, which blocked sunlight from reaching the robot's solar panels for weeks at a time.

Mars lies farther from the Sun than Earth does, so the Red Planet has a longer year – 687 days compared to 365 for our home world. But the two planets have similar day lengths; it takes about 24 hours and 40 minutes for Mars to complete one rotation around its axis. The axis of Mars, like Earth's, is tilted in relation to the Sun. This means that like Earth, the amount of sunlight falling on certain parts of the Red Planet can vary during the year, giving Mars seasons.

However, the seasons that Mars experiences are more extreme than Earth's because the Red Planet's elliptical, oval-shaped orbit is more elongated than that of any of the other major planets. When Mars is closest to the Sun, its southern hemisphere is tilted towards our star, giving the planet a short, warm summer, while the northern hemisphere experiences a short, cold winter. When Mars is farthest from the Sun, the northern hemisphere is tilted towards it, giving the planet a long, mild summer, while the southern hemisphere experiences a long, cold winter.

The tilt of the Red Planet's axis swings wildly over time because it's not stabilised by a large moon. This has led to different climates on the Martian surface throughout its history. A 2017 study suggests that the changing tilt also influenced the release of methane into

Perseverance records the Ingenuity helicopter's flights over the Martian surface for Earth viewing

UNDERSTANDING THE SOLAR SYSTEM

Measuring the vitals of Mars

Nothing beats the versatility of the InSight Mars lander. The mission is equipped with an array of sensors that have been described as 'taking the vitals' of Mars. They resulted in immediate success, with the detection of seismic activity – in the form of marsquakes – passing through the Red Planet.

Astronauts working safely on the surface of Mars with a pressurised vehicle in the background

Mars is thought to have a solid core like Earth

Atmosphere
Crust
Core
Mantle
Surface

Mars' atmosphere, causing temporary warming periods that allowed water to flow.

Mars is 6,791 kilometres (4,220 miles) in diameter – far smaller than Earth, which is 12,756 kilometres (7,926 miles) wide. The Red Planet is about ten per cent as massive as our home, with a gravitational pull 38 per cent as strong. A 45-kilogram person here on Earth would weigh just 28 kilograms on Mars, but their mass would be the same on both planets. The atmosphere of Mars is 95.32 per cent carbon dioxide, 2.7 per cent nitrogen, 1.6 per cent argon, 0.13 per cent oxygen and 0.08 per cent carbon monoxide, with minor amounts of water, nitrogen oxide, neon, heavy water, krypton and xenon.

Mars lost its global magnetic field about 4 billion years ago, leading to the stripping of much of its atmosphere by the solar wind. But there are regions of the planet's crust today that can be at least ten times more strongly magnetised than anything measured on Earth, suggesting those regions are remnants of an ancient global magnetic field. Mars likely has a solid core composed of iron, nickel and sulphur. The mantle is probably similar to Earth's in that it's composed mostly of peridotite, made up of silicon, oxygen, iron and magnesium. The crust is probably made of the volcanic rock basalt, which is also common in the crusts of Earth and the Moon, although some crustal rocks, especially in the northern hemisphere, may be a form of andesite, a volcanic rock that contains more silica than basalt does.

NASA's InSight lander has been probing the Martian interior since touching down near the planet's equator in November 2018. InSight measures and characterises marsquakes, and mission team members are tracking wobbles in Mars' tilt over time by precisely tracking the lander's position on the planet's surface. Data has revealed key insights about Mars' internal structure. InSight team members recently estimated that the planet's core is 1,780 to 2,080 kilometres (1,110 to 1,300 miles) wide. InSight's observations also suggest that Mars' crust is 24 to 72 kilometres (14 to 45 miles) thick on average, with the mantle making up the rest of the planet's non-atmospheric volume. For comparison, Earth's core is about 7,100 kilometres (4,400 miles) wide – bigger than Mars itself – and its mantle is roughly 2,900 kilometres (1,800 miles) thick. Earth has two kinds of crust, continental and oceanic. Average thicknesses are about 40 kilometres (25 miles) and eight kilometres (five miles) respectively.

The two moons of Mars, Phobos and Deimos, were discovered by American astronomer Asaph Hall over the course of a week in 1877. Hall had almost given up his search for a moon of Mars, but his wife urged him on. He discovered Deimos the next night, and Phobos six days after that. He named the moons after the sons of the Greek war god Ares – Phobos means 'fear', while Deimos means 'dread'. Both are likely made of carbon-rich rock mixed with ice and are covered in dust and loose rocks. They are tiny next to Earth's Moon and are irregularly shaped since they lack enough gravity to pull themselves into a more circular form. The widest Phobos gets is about 27 kilometres (17 miles), and the widest Deimos gets is roughly 15 kilometres (nine miles).

Both moons are pockmarked with craters from meteor impacts. The surface of Phobos also possesses an intricate pattern of grooves, which may be cracks that formed after an impact created the moon's largest crater – a hole about ten kilometres (six miles) wide, nearly half the width of Phobos. The two satellites always show the same face to their parent planet, just as our Moon does to Earth. It remains uncertain how Phobos and Deimos formed. They may be former asteroids that were captured by Mars' gravitational pull, or they may have formed in orbit around Mars at roughly the same time the planet came into existence. Ultraviolet light reflected from Phobos provides strong evidence that the moon is a captured asteroid. Phobos is gradually spiralling towards Mars, drawing about 1.8 metres (six feet) closer to the Red Planet each century. Within 50 million years, Phobos will either smash into Mars or break up and form a ring of debris around the planet.

The first person to observe Mars with a telescope was Galileo Galilei in 1610. In the following century, astronomers discovered the planet's polar ice caps. In the 19th and 20th centuries, some – most famously Percival Lowell – believed they saw a network of long, straight canals on Mars that hinted at a possible civilisation. However, these sightings proved to be mistaken interpretations of geological features.

A number of Martian rocks have fallen to Earth over the aeons, providing scientists with a rare opportunity to study pieces of Mars without having to leave our planet. One of the most controversial finds was Allan Hills 84001 (ALH84001) – a Martian meteorite that may contain tiny fossils and other evidence of Mars life. Other researchers have cast doubt on this hypothesis, but the team behind the famous 1996 study have held firm to their interpretation, and the debate about ALH84001 continues today. In 2018, a separate meteorite study found that organic molecules – the carbon-containing building blocks of life, although not necessarily evidence of life itself – could have formed on Mars through battery-like chemical reactions.

The Perseverance rover captures a picture as it selects a boulder to drill

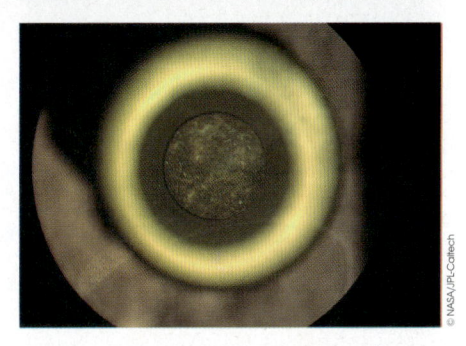

Robotic spacecraft began observing Mars in the 1960s, with the US launching Mariner 4 in 1964 and Mariners 6 and 7 in 1969. Those early missions revealed Mars to be a barren world, without any signs of life or the civilisations people such as Lowell had imagined there. In 1971, Mariner 9 orbited Mars, mapping about 80 per cent of the planet and discovering its volcanoes and canyons. The Soviet Union also launched numerous Red Planet spacecraft in the 1960s and early 1970s, but most of those missions failed. Mars 2 and Mars 3 operated successfully, but were unable to map the surface due to dust storms. NASA's Viking 1 lander touched down on the surface of Mars in 1976, pulling off the first successful landing on the Red Planet. Its twin, Viking 2, landed six weeks later in a different Mars region. The Viking landers took the first close-up pictures of the Martian surface, but found no strong evidence for life. Again, however, there has been debate: Gilbert Levin, principal investigator of the Vikings'

UNDERSTANDING THE SOLAR SYSTEM

Labeled Release life-detection experiment, forever maintained that the landers spied evidence of microbial metabolism in the Martian dirt.

The next two craft to successfully reach the Red Planet were Mars Pathfinder, a lander, and Mars Global Surveyor, an orbiter, both NASA craft that launched in 1996. A small robot on board Pathfinder, called Sojourner, became the first wheeled rover ever to explore the surface of another planet, venturing over the planet's surface and analysing rocks for 95 Earth days.

In 2001, NASA launched the Mars Odyssey orbiter, which discovered vast amounts of water ice beneath the Martian surface, mostly in the upper metre (three feet). It remains uncertain whether more water lies underneath, since the probe cannot see water any deeper.

In 2003, Mars passed closer to Earth than it had at any time in the past 60,000 years. That same year, NASA launched two golf-cart-sized rovers, named Spirit and Opportunity, which explored different regions of the Martian surface after touching down in January 2004. Both rovers found many signs that water once flowed on the surface. Spirit and Opportunity were originally tasked with three-month surface missions, but both kept roving for far longer than that. NASA didn't declare Spirit dead until 2011, and Opportunity was still going strong until that dust storm hit in mid-2018.

Next, in 2008, NASA sent a lander called Phoenix to the far-northern plains of Mars. The robot confirmed the presence of water ice in the near subsurface, among other finds. In 2011, NASA's Mars Science Laboratory mission sent the Curiosity rover to investigate Mars' past potential to host life. Not long after landing inside the Red Planet's Gale crater in August 2012, the car-sized robot determined that the area hosted a long-lived, potentially habitable lake-and-stream system in the ancient past.

Curiosity has also discovered complex organic molecules and documented seasonal fluctuations in methane concentrations in the atmosphere.

But NASA isn't the only interested party. The ESA has two spacecraft orbiting the planet: Mars Express and the Trace Gas Orbiter. Also, in September 2014, India's Mars Orbiter Mission reached the Red Planet, making it the fourth nation to successfully enter orbit around Mars.

In November 2018, NASA landed a stationary craft called InSight on the surface. InSight is investigating Mars' internal structure and composition, primarily by measuring and characterising marsquakes. NASA also launched the life-hunting Perseverance rover in July 2020. Perseverance landed on the floor of Mars' Jezero crater in February 2021 along with a tiny test helicopter known as Ingenuity.

2020 also saw the launch of the United Arab Emirates' first Mars mission, Hope, and China's first fully homegrown

"The Viking landers took the first close-up pictures of the Martian surface, but found no strong evidence for life"

MARS

Mars effort, Tianwen-1. Hope arrived in February 2021 and is studying the atmosphere, weather and climate. Tianwen-1, which consists of an orbiter and a lander-rover duo, also reached Mars orbit in February 2021.

Robots aren't the only ones getting a ticket to Mars. A workshop group of scientists from government agencies, academia and industry have determined that a NASA-led manned mission to Mars should be possible by the 2030s. Robotic missions have seen much success in the past few decades, but it remains a considerable challenge to get people to Mars. With current technology, it would take at least six months for people to travel to Mars. Red Planet explorers would therefore be exposed for long stretches to deep-space radiation and to microgravity, which has devastating effects on the human body. Performing activities in the moderate gravity of Mars could prove extremely difficult after many months in microgravity. Research into the effects of microgravity continues on the International Space Station in preparation.

NASA isn't the only entity with crewed Mars aspirations. Other nations, including China and Russia, have also announced their goals for sending humans to the Red Planet. And Elon Musk, the founder and CEO of SpaceX, has long stressed that he established the company back in 2002 primarily to help humanity settle the Red Planet. SpaceX is currently developing and testing a fully reusable deep-space transportation system called Starship, which Musk believes is the breakthrough needed to get people to Mars at long last.

NASA engineers working on a heat shield that could protect human visitors to Mars

Charles Q. Choi
Space science writer
Charles is a contributing writer for space.com and Live Science. He covers all things human origins and astronomy as well as physics.

Robert Lea
Space science writer
Rob is a writer with a degree in physics and astronomy. He specialises in physics, astronomy, astrophysics and quantum physics.

UNDERSTANDING THE SOLAR SYSTEM

How the planets would look...

...if they were at the same distance from Earth as the Moon, and how it would affect our planet

1 Mercury
Mercury is approximately 4.5 times more massive than the Moon. This would no doubt affect the tides on Earth. It also has a magnetic field that could affect Earth's.

2 The Moon
With a radius of 1,737.4 kilometres (1,079.6 miles), our Moon takes up about as much sky as the Sun, which is how they can appear to be the same size and distance away.

3 Venus
Venus has a very dense, toxic atmosphere, with a surface that can't be easily viewed. If it were closer, Earth's magnetic pull might affect that atmosphere, and vice versa.

4 Mars
If Mars were as close as our Moon, travelling there would be simpler. We could see many of its features with the naked eye, and a Martian colony might be possible.

5 Jupiter
It's unlikely that we'd survive if Jupiter became our Moon. It would blast us with its deadly radiation field and subject Earth to incredibly strong tidal stresses.

6 Saturn
Saturn's amazing rings would stretch nearly from horizon to horizon, and its banded atmosphere is calm in comparison to some of the other planets.

Expert: Robin Hague
Robin is a science writer, focusing on space and physics. He is head of launch at Skyrora, coordinating launch opportunities for Skyrora's vehicles.

HOW THE PLANETS WOULD LOOK...

What if the other Solar System planets were the same distance from Earth as the Moon? At best, a night sky with one or more planets at the Moon's distance would look very different. At worst, either the other planet or Earth – or both – would be radically changed by its proximity to the other. In some cases, we wouldn't survive. There are numerous factors involved, but speculation is part of the fun of astronomy.

Mercury, the closest planet to the Sun, is also a similar size to our Moon, with a radius of 2,439.7 kilometres (1,516 miles). It has similar surface features to the Moon from a distance, although if it were close up we'd be able to see that the craters, rays and other features are quite different.

Venus has a radius of 6,051.8 kilometres (3,760.4 miles) and would look as large to us as planet Earth looked to the Apollo astronauts when they were walking on the Moon. It reflects six times the amount of sunlight that the Moon does and would take up about 12 times the space in the sky from our perspective, so 'night', when Venus is on the opposite side of the sky from the Sun, would be much brighter.

Mars is about twice the size of the Moon, with a radius of 3,389.5 kilometres (2,106 miles), so it would look larger than the Sun does to us. Having Mars or any other planet larger than our Moon for a satellite would have an effect on the tides, causing huge waves and even tsunamis. Beach-going would probably be a thing of the past, but we'd have more light at night, and it would have a creepy red tint to it.

Jupiter would completely dominate our sky. Astronomers measure the distances between objects in the sky using degrees – our Moon takes up about half a degree, but Jupiter would take up 20 degrees. We wouldn't be able to see Jupiter's poles, but we would be treated to a view of its distinctive caramel-and-white bands and the ongoing storms on the planet. Saturn has a radius that's more than nine times that of Earth's, and in truth Earth would really become a satellite of Saturn instead.

Although not as huge as Jupiter or Saturn, Neptune is still 14 times larger than the Moon. This big, blue planet would still dominate the sky at all times. There's a reason why Neptune and Uranus are sometimes called sister planets: both are icy worlds and have a radius about four times that of Earth's, although Uranus has a much calmer atmosphere than the other turbulent giant planets.

7 Uranus
One of Uranus' unique features as a Moon would be the fact that it rotates on its 'side' – more like a ball rolling around instead of a top spinning.

8 Neptune
Neptune has a very active atmosphere with lots of storms and other meteorological activity, which would provide us with compelling sights from Earth.

9 Pluto
Pluto's a dwarf planet with a diameter of 2,368 kilometres (1,471 miles) – around two-thirds the size of our Moon with less than half our Moon's gravity.

UNDERSTANDING THE SOLAR SYSTEM

Jupiter

The largest planet has a lot to tell us, and Juno is on the case

Fifth in the eight-planet lineup of our Solar System, Jupiter also happens to be the largest, and by quite some distance. The mass of this gigantic ball of gas is two-and-a-half times that of all the other planets put together, and you could fit 11.2 Earths within its radius. While there's likely a rocky core somewhere under the enormous gaseous atmosphere, scientists can't be sure whether it's solid or not, but gravitational measurements suggest it could make up as much as 15 per cent of Jupiter's mass.

What is known is that Jupiter is contracting, and this generates more heat than the planet receives from the Sun, warming the huge number of moons that orbit around it. It also has a faint ring system – too thin to be seen from Earth with any but the largest telescopes and first spotted by the Voyager 1 probe in 1979.

Jupiter plays a major role in many theories of the formation of our Solar System. In the grand tack hypothesis, Jupiter formed at 3.5 astronomical units (AU) – 1 AU is the Earth-Sun distance – before plunging inward towards the Sun until it reached 1.5 AU, then reversing course and moving out again, stopping at its current distance of 5.2 AU. It crossed the asteroid belt twice, scattering rocks in all directions and contributing to the low mass of the belt today. It may also have caused rocky planets orbiting closer to the Sun to crash into the star's surface. This answers questions such as why Mars is so small – Jupiter's presence limited the material available for its formation – and why there are no large planets orbiting close to the Sun, as we see in other solar systems.

Jupiter has also had a long-lasting effect on the rest of the Solar System. It has a fleet of asteroids and comets that follow it through its orbit – over 2,000 have been discovered – and its great mass means that the centre of gravity for it and the Sun lies above the Sun's surface, meaning they act almost like a binary system. The giant planet's gravity well also means it can intercept comets and asteroids heading into the inner Solar System and may partially shield the inner planets from bombardment. Another theory is that it draws small bodies in from the Kuiper Belt. Whichever is true, Jupiter experiences 200 times more impacts than Earth.

Galileo discovered Jupiter's four largest moons, known as the Galilean moons, in 1610 – the first time moons had been observed around another planet. Humanity has since explored the planet with observatories and space probes, beginning in 1973 with a flyby by Pioneer 10. Many missions to the outer Solar System have used Jupiter's gravity as a slingshot to correct their course or gain speed, but the first craft to orbit the planet was the aptly named Galileo in 1995.

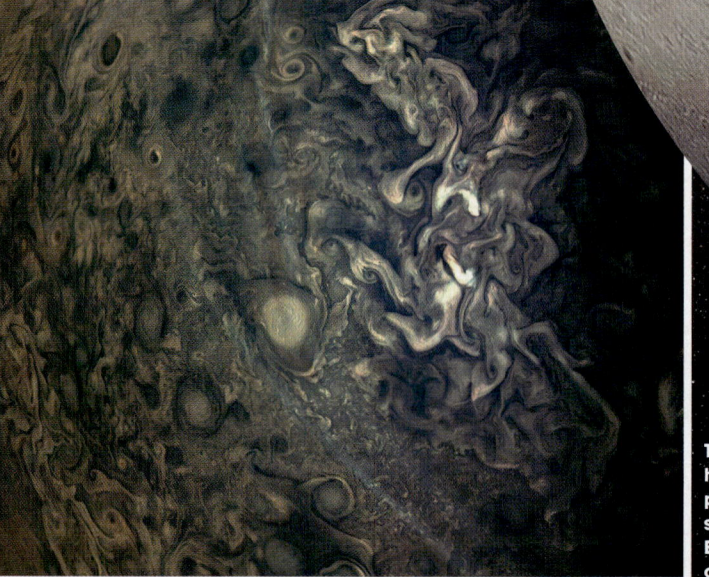

The hazy northern hemisphere of Jupiter processed by citizen scientist Gerald Eichstädt from Juno camera data in 2020

JUPITER

A cyclonic storm in
Jupiter's northern
hemisphere, taken by
Juno in 2019

Atmospheric composition

Upper atmosphere

90%
Hydrogen

10%
Helium

Lower atmosphere

71%
Hydrogen

24%
Helium

5%
other
elements

51

UNDERSTANDING THE SOLAR SYSTEM

News from Jupiter
Delve into the Gas Giant's fascinating secrets

Wind speeds measured

Scientists have directly measured the winds in the middle of Jupiter's atmosphere. Using the Atacama Large Millimeter/submillimeter Array (ALMA), a team was able to track the movement of molecules of hydrogen cyanide in the planet's turbulent atmosphere, measuring narrow bands of wind at up to 1,448 kilometres (900 miles) per hour. Hydrogen cyanide is not native to Jupiter, but was added when Comet Shoemaker-Levy 9 collided with the planet in 1994. Since then it has been circling the atmosphere. Using 42 of ALMA's 66 high-precision antennae, a team measured the Doppler shift, tiny changes in the radiation emitted by the molecules, from which they were able to deduce wind speed.

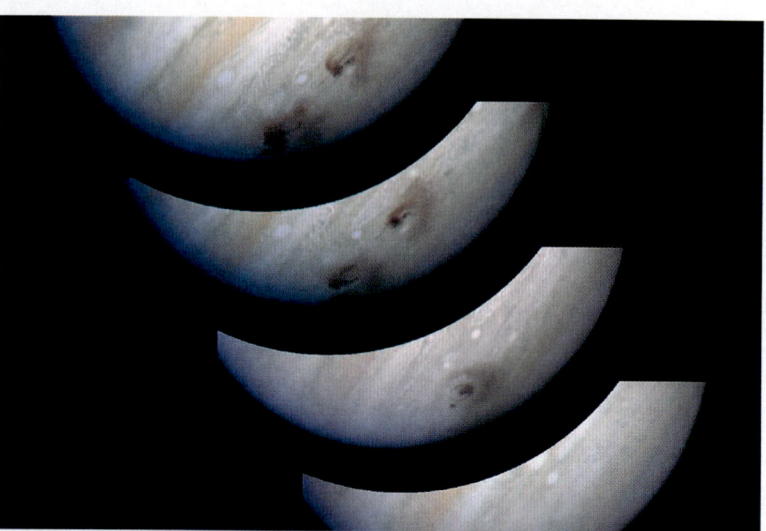

Auroral activity

Jupiter's version of the northern lights has puzzled scientists because it doesn't behave like aurorae on Earth. Here the lights appear in a ring between 60 and 70 degrees north or south of the equator. Within that ring – an area known as the 'polar cap' – they don't appear. On Jupiter there is no 'polar cap', so aurorae continue all the way to the poles. This is due to a quirk of Jupiter's magnetic field. On Earth, the aurorae appear on closed field lines, which extend outwards from the planet before bending back again. Inside the 'polar cap' the field lines are open and there are no aurorae. But Jupiter has a mixture of open and closed field lines as you approach its poles, meaning the aurorae are still able to appear.

Another Jupiter

Little is known about how planets as large as Jupiter form, but a planet circling another star – and under the watchful eye of Hubble – could give us a lot of information. Known as PDS 70b, the planet orbits a very young orange dwarf 370 light years away in the southern constellation of Centaurus, which has two actively forming planets within its protoplanetary disc. PDS 70b, which orbits the star at the same distance as Uranus orbits our Sun, is already around five times the mass of Jupiter – and possibly twice as large – and at a mere 5 million years old should continue to form for a little while yet, though the rate at which it is accreting more material has dwindled.

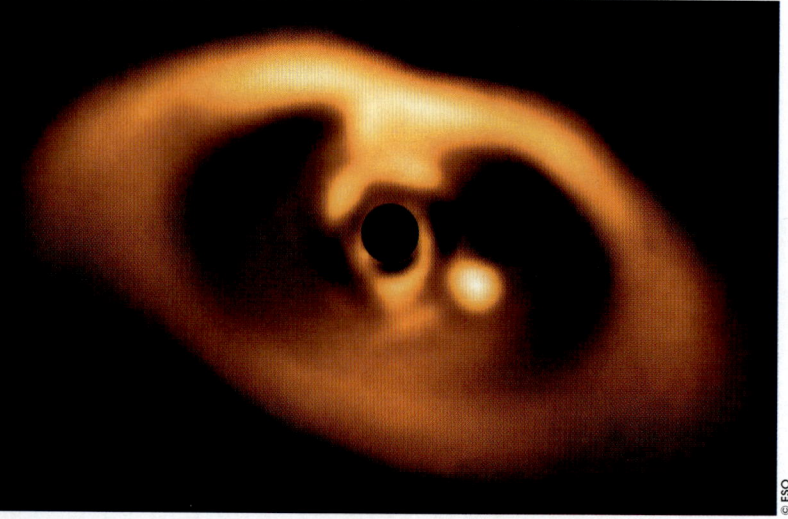

52

Jupiter facts

14x — Jupiter's magnetic field is 14 times stronger than Earth's and is the strongest in the Solar System except for sunspots

5,000 kilometres — The thickness of Jupiter's atmosphere – the deepest in the Solar System

1665 — Jupiter's Great Red Spot is a storm known to have existed since at least 1831, and maybe even since 1665

11.8 — Years Jupiter takes to orbit the Sun

4th — Fourth-brightest object in the sky as seen from Earth

79 — known moons circulate around Jupiter

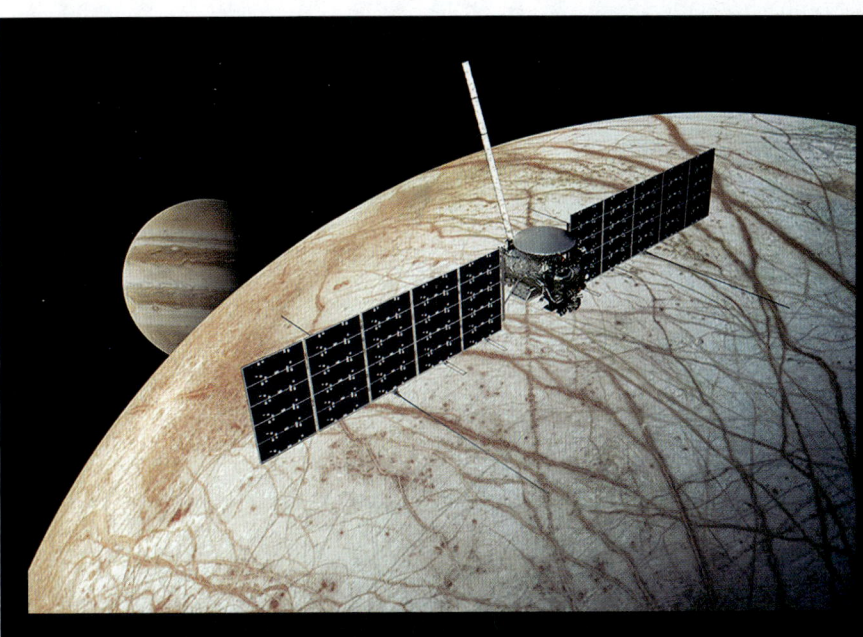

Evolution of the Jovian giant

Time: 4.6 billion years ago
Activity: The Solar System began to form from a cloud of gas and dust around a new star.

Time: 4.596 billion years ago
Activity: Jupiter and Saturn began to take shape.

Time: 2400 BCE
Activity: Babylonians tracked a full cycle of Jupiter's movement across the skies.

Time: 270 BCE
Activity: Jupiter was part of Aristarchus of Samos' heliocentric model of the Solar System.

Time: 1610
Activity: Galileo discovered the Galilean moons: Ganymede, Callisto, Io and Europa.

Time: 3 December 1974
Activity: Pioneer 11 passed within 42,500 kilometres (26,400 miles) of Jupiter's cloud tops.

Time: 5 March 1979
Activity: Voyager 1 performed a flyby of the gas giant planet.

Time: 8 December 1995
Activity: The Galileo probe entered Jupiter orbit.

Time: 5 July 2016
Activity: The Juno probe entered a polar orbit around the planet.

Future plans for Jupiter

While Jupiter has been heavily photographed by missions such as Juno, which arrived at the planet in 2016, much scientific interest has now transferred to the planet's moons, which are thought to harbour subsurface liquid oceans and possibly even life. Europa, Ganymede and Callisto, three of the Galilean moons, would be the targets, but multiple missions have been cancelled due to lack of budget.

In 2024 NASA's Europa Clipper should launch, following up on studies from the Galileo probe and performing multiple flybys of Europa without orbiting it, using the gravity of nearby moons to change its course. The European Space Agency sent its Jupiter Icy Moons Explorer in April 2023 to study Ganymede, Callisto and Europa, to evaluate their potential to support life. Other countries also have their eyes on the giant planet, with China's Gan De craft proposed for launch in 2029 and an unnamed Russian proposal to use a nuclear-powered tug to travel to the planet sometime after 2030.

Further into the future, Europa is seen as a potential site for human colonisation of the Solar System, as it is geologically stable and levels of radiation are low there. Low is a relative term, however, as unshielded colonists would receive 5.4 sieverts of radiation per day from Jupiter compared to 0.0024 sieverts per year on Earth. This is still enough to cause radiation poisoning.

UNDERSTANDING THE SOLAR SYSTEM

Everything you need to know about Saturn

With its moon system, rings of dust and ice and occasionally tempestuous atmosphere, there's more to this gas giant than meets the eye

SATURN

Saturn is our Solar System's ringed wonder – a spectacular world encircled by planes of icy debris, giving it a unique appearance. But there's a lot more to Saturn than just its rings; this enormous world is worth exploring both for its own complexity and the fascinating family of satellites that orbit it. As the most distant Solar System object easily seen with the naked eye, Saturn orbits at an average of 1.43 billion kilometres (887 million miles) from the Sun. Its slow orbit means that Saturn takes 29.5 years to make a full circuit through the constellations of the zodiac; it was this stately movement that led ancient stargazers to associate it with the father of Jupiter in Roman mythology.

Its distance makes it a challenging object for study, even in the era of giant telescopes. Most of what we know about the planet comes from the Voyager probe flybys in the 1980s and the Cassini mission that orbited between 2004 and 2017. Earth observations, coupled with close-up images from these explorers, have revealed that what often appears to be a placid orb of creamy cloud is in fact a surprisingly active world.

Internally, Saturn is a gas giant like Jupiter, a huge ball dominated by the lightweight elements hydrogen and helium. It owes its very different outward appearance to a substantially lower mass – Saturn weighs as much as 95 Earths,

Saturn's Moons

Daphnis
Dimensions: 8.6 by 8.2 by 6.4 kilometres (5.3 by 5.1 by 4.0 miles)
Mass: 0.08 x 10^{12} tonnes
Orbital period: 0.59 days
Discovered: Cassini, 2005

Mimas
Diameter: 396 kilometres (246 miles)
Mass: 37.4 x 10^{15} tonnes
Orbital period: 0.94 days
Discovered: William Herschel, 1789

Enceladus
Diameter: 504 kilometres (313 miles)
Mass: 108.0 x 10^{15} tonnes
Orbital period: 1.37 days
Discovered: William Herschel, 1789

Tethys
Diameter: 1,062 kilometres (659 miles)
Mass: 617.4 x 10^{15} tonnes
Orbital period: 1.89 days
Discovered: Giovanni Domenico Cassini, 1684

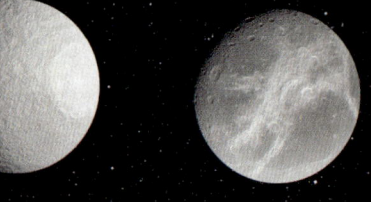

Dione
Diameter: 1,123 kilometres (698 miles)
Mass: 1095.4 x 10^{15} tonnes
Orbital period: 2.74 days
Discovered: Giovanni Domenico Cassini, 1684

---

but this is less than a third of Jupiter. The weaker gravity allows Saturn's upper layers to billow outwards, giving it the lowest average density of any world in the Solar System – about two-thirds that of water.

Combined with a rotation period of just ten-and-a-half hours, Saturn struggles to hold onto material around its fast-moving equator, giving the planet a pronounced bulge around the middle. At 120,532 kilometres (74,895 miles), Saturn's equatorial diameter is almost 12,000 kilometres (7,456 miles) more than its polar diameter. The planet's low density also means its atmosphere is cooler, and combined with its distance from the Sun means that temperatures in the uppermost layers plunge to a chilly -176 degrees Celsius (-285 degrees Fahrenheit). This creates conditions where ammonia can condense into ice crystals and form hazy white clouds that blanket the planet, muting the colours and detail of features deeper inside the atmosphere. Visiting space probes have confirmed a wide range of atmospheric compounds, including ethane, methane and ammonia, while the deeper clouds likely consist of ammonium hydrosulfide or water. Differences in the cocktail of chemicals are probably mostly responsible for Saturn's restrained palette in comparison with Jupiter.

Despite its relatively monotone appearance, Saturn's atmosphere is far from dormant. Space probes have recorded wind speeds of up to 1,800 kilometres (1,118 miles) per hour, the second fastest in the Solar System after Neptune, forming jet streams that wrap around the planet. Saturn's major cloud bands run parallel to the equator; they are broader and fewer in number than those around Jupiter, but still give rise to intense storms that appear as bright ovals and persist for weeks.

Although Saturn has no permanently visible storms to match Jupiter's Great Red Spot, it intermittently produces seasonal storms on just as grand a scale. The most famous, known as the Great White Spot, erupted in the northern hemisphere at roughly 30-year intervals from 1876 until 1990. Its regular cycle suggests a link to seasonal changes in the upper atmosphere. With its equator tipped at an angle of 26.7 degrees to its orbit, Saturn goes through an Earthlike pattern of seasons each long year. But the spot's unexpected early reappearance in 2010 – and absence since – suggest things may be more complicated.

Beneath its outer cloud layers, Saturn's internal structure is governed by the increasing temperature and atmospheric pressure deeper inside the planet. About 1,000 kilometres (621 miles) below the visible surface, its hydrogen gas is so compressed that it condenses into liquid. Some way below this, conditions become so extreme that hydrogen molecules are broken apart, forming a sea of electrically charged liquid metallic hydrogen that generates a powerful magnetic field as the planet spins. The exact structure of Saturn's core remains something of a

Taken as Cassini passed through Saturn's shadow, this spectacular image reveals the planet and its ring system backlit by the Sun

SATURN

Rhea
Diameter: 1,528 kilometres (949 miles)
Mass: 2306.5 x 10^{15} tonnes
Orbital period: 4.52 days
Discovered: Giovanni Domenico Cassini, 1672

Titan
Diameter: 5,149 kilometres (3,199 miles)
Mass: 134,520.0 x 10^{15} tonnes
Orbital period: 15.95 days
Discovered: Christiaan Huygens, 1655

Hyperion
Dimensions: 360 by 266 by 205 kilometres (223 by 165 by 127 miles)
Mass: 5.6 x 10^{15} tonnes
Orbital period: 21.28 days
Discovered: William Cranch Bond, William Lassell, 1848

Iapetus
Diameter: 1,469 kilometres (913 miles)
Mass: 1805.6 x 10^{15} tonnes
Orbital period: 79.32 days
Discovered: Giovanni Domenico Cassini, 1671

Phoebe
Dimensions: 219 by 217 by 204 kilometres (136 by 134 by 127 miles)
Mass: 8.3 x 10^{15} tonnes
Orbital period: 550 days
Discovered: William Henry Pickering, 1899

---

mystery, but recent studies of the way it affects the planet's rings suggests it is probably fuzzy or diffuse, amounting to about 17 Earth masses of material and extending 60 per cent of the way to the surface, where it becomes mixed with the metallic hydrogen above. Deep within the planet, one or more mechanisms generate vast amounts of heat, allowing Saturn to radiate 2.5 times more energy than it receives from the Sun. In part this is probably due to a well-understood mechanism of gravitational contraction that sifts denser material towards the core, but it's likely that other chemical and physical processes within the liquid layers also play a role.

Although all four of the Solar System's giant planets are now known to have rings of one kind or another, Saturn's are by far the brightest and most extensive. They were first spotted by Galileo Galilei in 1610, though his crude telescope could only show that the planet appeared strangely elongated. It wasn't until 1659 that Christiaan Huygens realised the true structure of the rings as a broad but thin disc around the planet, which disappears from view when the plane of the rings lines up with Earth twice in each Saturnian year.

The two brightest rings, stretching to about 2.4 times Saturn's diameter, are designated A and B, and are separated by a mostly empty gap, the Cassini Division. Inward of the B Ring lies the C Ring, semi-transparent but still visible from Earth,

Inside Saturn

1 Ammonia haze
Cold conditions in the upper atmosphere allow ammonia to condense and form bright, hazy clouds at pressures between roughly 0.4 and 1.7 times Earth's atmospheric pressure.

2 Deeper clouds
As pressure increases further into Saturn's atmosphere, other chemicals condense into droplets and form clouds – in particular ammonium hydrosulfide and a water ice/ammonia mix.

3 Liquefied gas
At a depth of about 1,000 kilometres (621 miles), pressures reach 1,000 Earth atmospheres – enough for hydrogen to condense into its liquid molecular form.

4 Metallic ocean
Here pressures reach 2 million Earth atmospheres and temperatures rival the surface of the Sun. Hydrogen molecules split apart to form a sea of electrically charged metallic hydrogen.

5 Fuzzy core
Recent studies suggest Saturn has a fuzzy core that begins where heavy elements mix with the liquid metallic layer, and it grows denser towards the centre.

6 Solid centre?
The scientific jury is still out on whether Saturn has a distinct solid core of rock and metal.

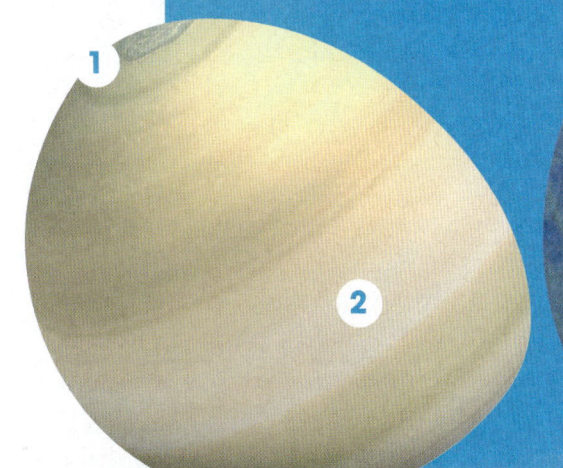

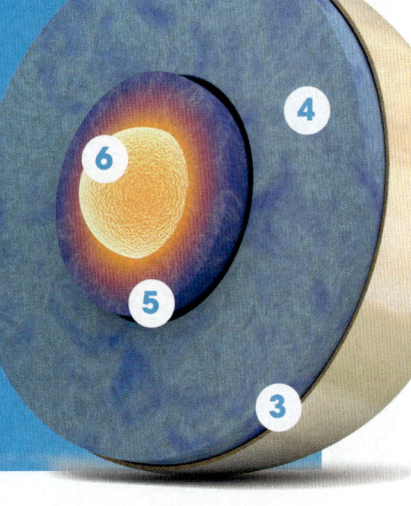

UNDERSTANDING THE SOLAR SYSTEM

After reappearing in late 2010, Saturn's Great White Spot soon developed a tail that wrapped around the planet as it became distorted by high winds

while a tenuous D Ring extends all the way down to Saturn's upper atmosphere. The narrow braid of the F Ring hems the A Ring's outer edge, while several minor rings – fragmentary 'arcs' or clouds of tiny particles – lie still further out.

The laws of physics mean that it's impossible for rings to be solid bodies – the varying strength of Saturn's gravity would tear them apart. Instead, each ring – and the many distinct ringlets within them – is composed of countless icy fragments, each following its own individual orbit around the planet. The orbits are almost perfectly circular and lie in a single narrow plane less than a kilometre (0.62 miles) deep, but the system is constantly evolving as the gravity of Saturn's major moons pulls fragments out of this neat arrangement and collisions with their neighbours jostle them back into line. Cassini images have revealed increasing levels of detail within the rings, including dark radial 'spokes' rippling their way across the rings like a wave and propeller-shaped structures caused by ring fragments clumping together.

Within and just beyond the rings, probes have traced the influence of so-called 'shepherd moons' such as Daphnis – small satellites ranging from a few kilometres to a few tens of kilometres across that keep individual ringlets in line, create gaps in the system and may even contribute fresh material to the rings. The evolving nature of the system, which grinds down ring fragments over time and ultimately loses material as it spirals down onto Saturn's equator, means that the rings must either be relatively young – one recent analysis points to formation from the breakup of a large, icy comet about 100 million years ago – or have somehow been replenished with new material over time.

The shepherd moons are just the innermost members of a vast family; with 82 members at the most recent count, Saturn has the biggest satellite system of any of the major planets. Beyond the rings orbit eight major moons, substantial worlds that formed from material left behind during the birth of Saturn itself. The inner five major moons follow a broad trend of increasing size, from 400-kilometre (248-mile) Mimas, through Enceladus, Tethys and Dione, to 1,500-kilometre (932-mile) Rhea; each consists of a mix of ice and rock, and most show signs of geological activity at various stages in their past, most likely low-temperature cryovolcanism creating eruptions of icy slush. The exception is the extraordinary Enceladus – a brilliant-white world 504 kilometres (313 miles) in diameter whose terrain has not only been reshaped by very recent resurfacing, but is also blanketed in fresh snow. Heat generated as Enceladus is distorted by a gravitational tug of war between Saturn

Major missions to Saturn

Pioneer 11
Type: Jupiter and Saturn flyby
Launched: 5 April 1973
Saturn flyby: 1 September 1979
Key discoveries: First close-up images of Saturn. Measurements of the magnetic field. Measured the temperature of Titan
Status: Contact lost; now entering interstellar space

Voyager 1
Type: Jupiter and Saturn flyby
Launched: 5 September 1977
Saturn flyby: 12 November 1980
Key discoveries: Analysis of Saturn's upper atmosphere. Discovery of complex structure in the rings of Saturn
Status: Entered interstellar space in 2012; still returning data to Earth

Voyager 2
Type: Multi-planet flyby
Launched: 20 August 1977
Saturn flyby: 26 August 1981
Key discoveries: Measurement of temperature in the upper atmosphere. Images of more structure in the rings
Status: Entered interstellar space in 2018; still returning data to Earth

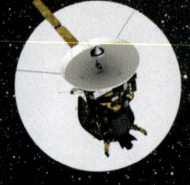

Cassini
Type: Saturn orbiter
Launched: 15 October 1997
Saturn orbit: 1 July 2004
Key discoveries: Detailed survey of Saturn, its rings and moons. Monitored Great White Spot outbreak of 2010. Imaged Titan through its clouds
Status: Destroyed during a controlled plunge into Saturn's atmosphere in 2017

Huygens
Type: Titan lander
Launched: 15 October 1997
Titan descent: 14 January 2005
Key discoveries: First descent through Titan's atmosphere and images from the surface. Landed in a dry river delta amid pebbles of water ice
Status: Contact with Cassini lost soon after landing

Dragonfly
Type: Planned Titan lander
Launch: 2027
Titan landing: 2034
Key goals: Dragonfly is a helicopter-like robotic explorer designed to fly across Titan's surface and visit several locations, studying the moon's complex surface chemistry and potential for life.

SATURN

Top 5 discoveries

1 The Dragon Storm
Usually hidden deep beneath the clouds, this long-lived storm occasionally sends plumes of white cloud to the surface and goes through periods of violent lightning activity.

2 Polar hexagon
Saturn's north pole is home to a giant hexagonal cloud structure 30,000 kilometres (18,641 miles) across. At its very centre lies a swirling vortex-like feature that's matched by one at the south pole.

3 Ring ripples
Saturn's rings are constantly reshaped by the gravitational influence of its moons – here the 86-kilometre (53-mile) Prometheus creates dark channels in the narrow F Ring as it disrupts the orbits of individual ring fragments.

4 Plumes of Enceladus
Jets of water vapour erupt from seas just beneath the icy crust of Enceladus, spraying snow across the surface and forming the doughnut-shaped E Ring that surrounds the moon's orbit.

5 Titan's eerie geography
Infrared images have revealed that Titan is a complex world with a fresh, mostly uncratered terrain.

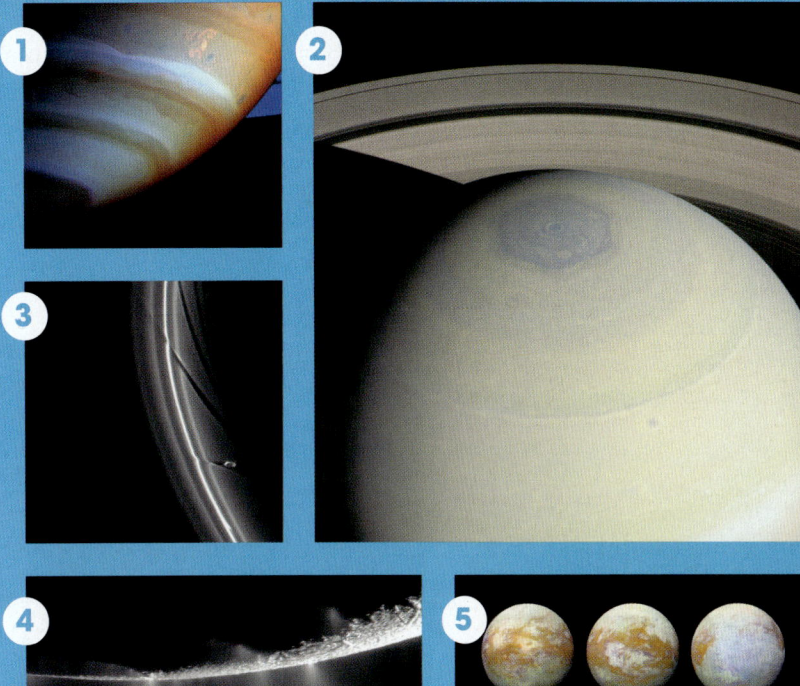

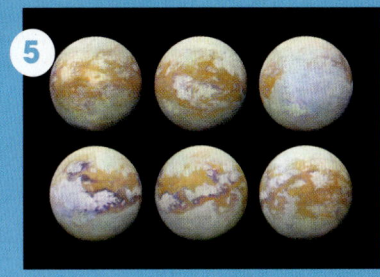

---

and its outer neighbouring satellites warms the moon's south polar region, creating reservoirs of liquid water just beneath the icy crust. As the surface flexes, water escapes into space, shooting huge plumes of vapour high above the moon's surface. Enceladus' near-surface liquid water makes it a prime candidate in the search for life elsewhere in the Solar System.

The centrepiece of Saturn's moon system, the appropriately named Titan, dwarfs all the other satellites. With a diameter of 5,149 kilometres (3,199 miles), it's the second-largest moon in the entire Solar System after Jupiter's Ganymede. A combination of high gravity and cold conditions allow Titan to hold onto a substantial atmosphere, making it one of a kind among Solar System satellites. This atmosphere is mostly nitrogen, but a small amount of methane forms clouds that render it an opaque orange. When the Cassini probe's infrared cameras pierced this veil, they revealed a curiously Earthlike landscape with eroded 'continents' and low-lying plains that resemble ocean basins. With an average temperature of -179 degrees Celsius (-111 degrees Fahrenheit), Titan's terrain has been shaped by both cryovolcanic activity and a 'methane cycle' resembling Earth's water cycle, This involves the volatile chemical shifting between atmospheric gas, solid ice and liquid rain that erodes the surface and gathers in lakes around the moon's winter pole.

The final two large moons are also intriguing. Spongelike and misshapen Hyperion is a mere 360 kilometres (223 miles) long, and is thought to be a surviving fragment of a moon that was once much larger. Iapetus, which rivals Rhea in size, has starkly contrasting light and dark hemispheres, the end result of a complex process that begins with Iapetus picking up dust that spirals Saturnwards from its dark outer neighbour Phoebe.

Phoebe itself is the largest of more than 50 outer 'irregular' moons – small icy bodies that follow tilted, elongated and sometimes backward orbits. Grouped into several distinct families, they are thought to be the fragmented remains of comets or asteroids from the outer Solar System, captured by Saturn's gravity long after their formation. Together, the outer limits of their distant orbits extend the limits of this dazzling system to more than 30 million kilometres (18.6 million miles) from Saturn itself.

Giles Sparrow
Space science writer

Giles has degrees in astronomy and science communication and has written many books and articles on all aspects of the universe.

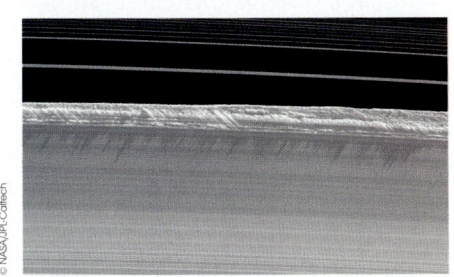

Cassini captured long shadows cast by ragged material at the outer edge of the B Ring as the Sun rose over the southern face of the rings shortly after Saturn's 2009 equinox

UNDERSTANDING THE SOLAR SYSTEM

Secrets of the Ice Giants

At the far reaches of the Solar System, two worlds remain a mystery to us. But recent research may have given us a peek into the unknown

SECRETS OF
THE ICE GIANTS

Uranus and Neptune could be stranger than we once thought. On the surface these two planets – roughly midway in size between Earth and Jupiter – seem unassuming, if different from one another. Uranus is a featureless, pale-azure planet, and Neptune a deep-blue one with white cloud bands and a dark storm system similar to Jupiter's Great Red Spot. But at heart they may be much more alike, as well as unlike anything we would encounter on Earth. Studies are showing that in terms of chemistry, density, temperature and pressure, the interiors of these worlds have the complexity of a Shakespearean character, and even that they may have actual diamond rain.

Although NASA's pioneering Voyager 2 spacecraft visited both planets in 1986 and 1989, sending back a wealth of images and data, no spacecraft has been to either world since. That may be due to the fact that Uranus and Neptune are the last two official planets, lying at the planetary edges of our Solar System 2.9 and 4.5 billion kilometres (1.8 and 2.8 billion miles) from the Sun respectively. Much of our information – and all of our up-close images of these worlds – come from Voyager 2, although both are studied by ground and space-based telescopes.

The giant planets formed in the outer Solar System where hydrogen and helium were more abundant. Clearly Uranus and Neptune aren't small and rocky like the planets of the inner Solar System. But nor do they quite reach the status of 'gas giant' like Jupiter and Saturn, even though they have similar bulk compositions of hydrogen and helium by percentage. They belong to a class of their own: so-called 'ice giants'. The 'ice' refers to some of the volatile chemicals found – mostly – deep within.

One scientist, Professor Jonathan Fortney of the University of California, Santa Cruz, says: "Uranus and Neptune are generally thought of as 'failed' versions of Jupiter and Saturn. They did not accrete tens to hundreds of Earth masses of hydrogen and helium, probably because there was less gas farther from the Sun by the time they formed." Fortney is a member of the science team for a proposed future mission to the ice giants led by NASA's Jet Propulsion Laboratory, Goddard Space Flight Center and the European Space Agency.

Hydrogen and helium would also have evaporated close to the Sun once it started shining. This volatility also applies to the ices, which include ammonia, water and methane compounds. Water might seem like a surprising addition, but it's liquid on Earth because of the pressure of our atmosphere, if you discount temperature variations.

The current broad consensus by planetary scientists is that both planets have rocky, iron-nickel Mars to Earth-sized cores; fluid, icy mantles that are 10 to 15 times Earth's mass – with Uranus' calculated to be 13.4 times – and hydrogen-

> "Uranus and Neptune are generally thought of as 'failed' versions of Jupiter and Saturn"

82.5% Hydrogen
15.2% Helium
2.3% Methane

Inside Uranus

1 Hydrogen-helium atmosphere
Although by percentage the atmospheric composition of hydrogen and helium is similar to the other giant planets, by mass it's very low.

2 Silicate iron-nickel core
The core is thought to be a rocky, iron-nickel body 0.5 to 3.7 times Earth's mass. The precise figure is unknown due to the difficulty of calculating it.

3 Fluid icy mantle
Made of ammonia, water and methane ices, the mantle is the bulk of Uranus' mass, increasing in temperature and pressure towards the core

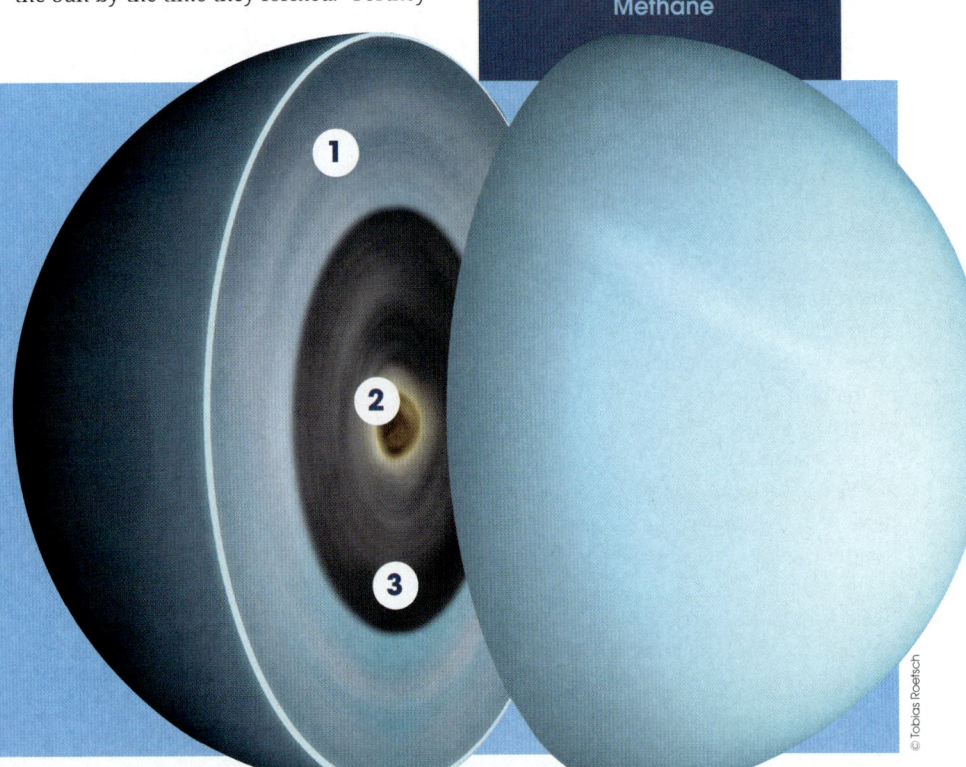

61

UNDERSTANDING THE SOLAR SYSTEM

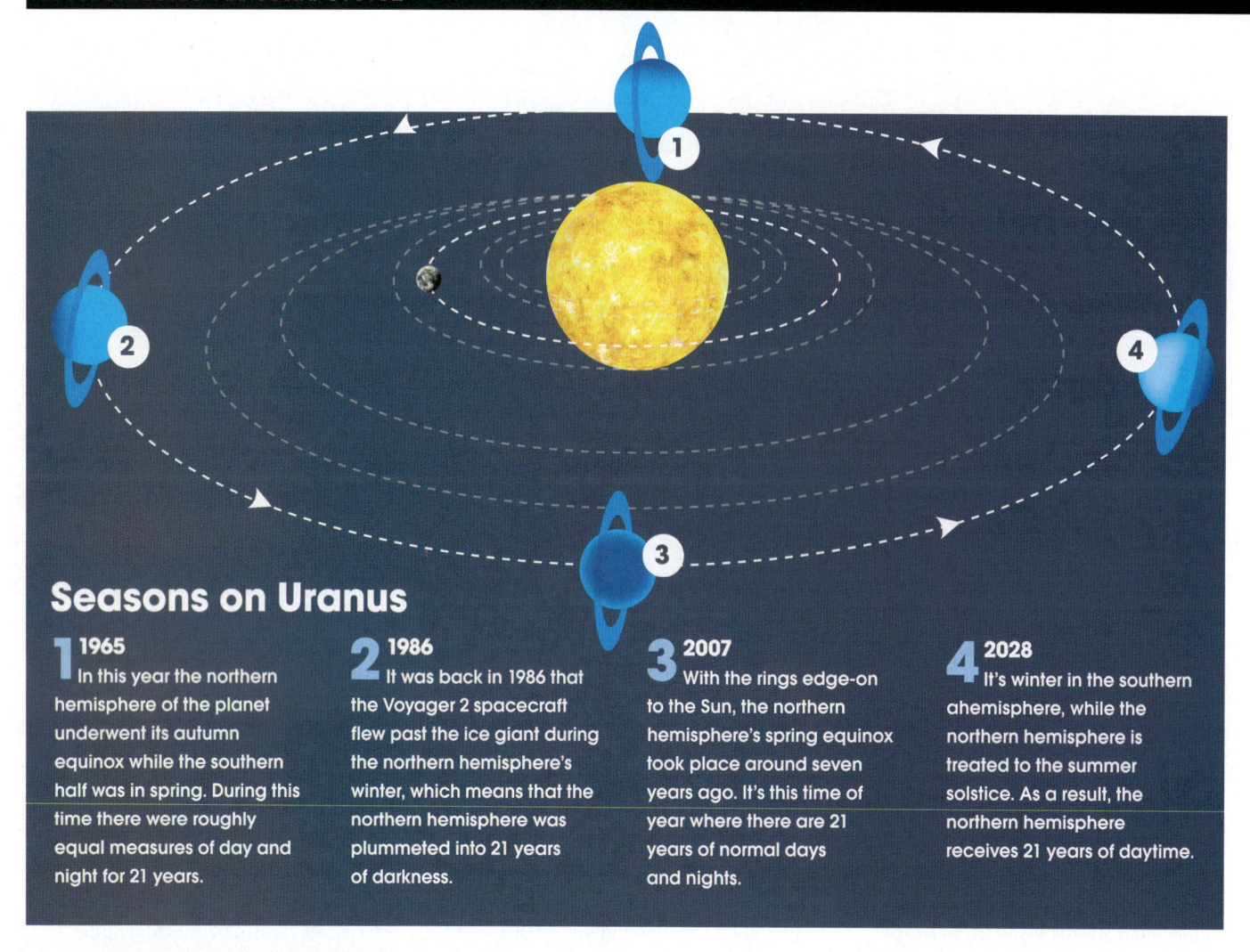

Seasons on Uranus

1 **1965**
In this year the northern hemisphere of the planet underwent its autumn equinox while the southern half was in spring. During this time there were roughly equal measures of day and night for 21 years.

2 **1986**
It was back in 1986 that the Voyager 2 spacecraft flew past the ice giant during the northern hemisphere's winter, which means that the northern hemisphere was plummeted into 21 years of darkness.

3 **2007**
With the rings edge-on to the Sun, the northern hemisphere's spring equinox took place around seven years ago. It's this time of year where there are 21 years of normal days and nights.

4 **2028**
It's winter in the southern ahemisphere, while the northern hemisphere is treated to the summer solstice. As a result, the northern hemisphere receives 21 years of daytime.

helium atmospheres with small amounts of methane. But even with the Voyager 2 data, scientists still don't really know what these worlds are like inside. Work done over the years almost always has to rely on computer simulations because the planets' internal conditions are so difficult to recreate with current laboratory equipment. Ices comprise the bulk mass of the ice giants' mantles, and their temperatures and pressures change at different altitudes. This is where the term 'ices' becomes weird in the conventional sense, because the ammonia, water and methane mixtures can reach thousands of degrees Celsius in temperature the further you go down. The reason they are able to maintain their composition is because of the soul-crushing pressures of hundreds of thousands to even millions of Earth atmospheres.

A recent study has investigated this further. An international team led by Dr Andreas Hermann of the University of Edinburgh's School of Physics and Astronomy and Centre for Science at Extreme Conditions simulated water and ammonia mixtures at low temperature conditions. The team discovered that the mixture allowed a compound called ammonia hemihydrate to remain stable as it went through ionic phases at increasingly high pressures. "There's an interesting state of matter – called superionic – where, due to heat, protons become diffuse, while the heavy atoms of carbon, nitrogen and oxygen remain in a fixed lattice," says Hermann. "It's a partially molten state that still carries signatures from the underlying crystal structures."

What this means is that molecules of superionic matter start off as a 'sea' of free-floating ions of their original selves, but then, under increasing pressure, they crystallise to become a strange liquid-

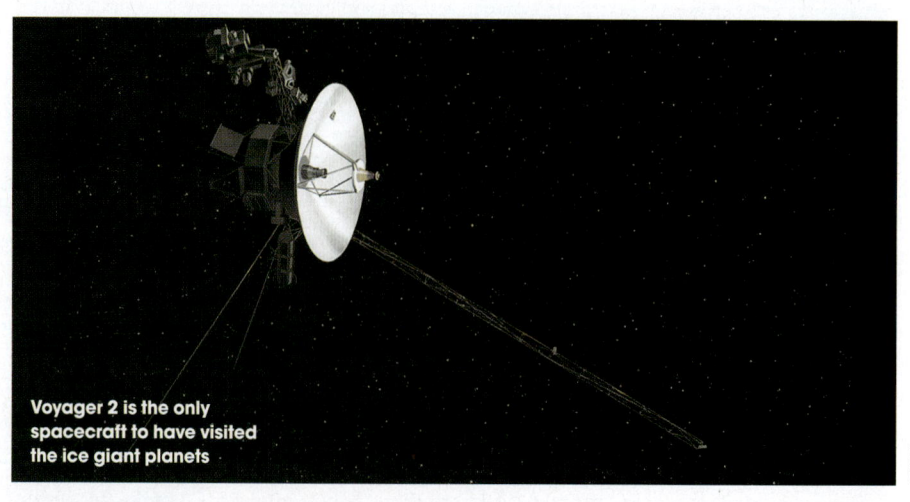

Voyager 2 is the only spacecraft to have visited the ice giant planets

62

SECRETS OF
THE ICE GIANTS

How an ice giant is made
The ice giants formed around the Sun like their Solar System siblings

1. Cloud of gas and dust collapses
An interstellar cloud of gas and dust, known as a solar nebula, collapsed in on itself and began to spin. Our newborn Sun began to shine in the centre of this spinning disc as the temperatures and pressures triggered thermonuclear fusion.

2. Material comes together
Heavier materials in the spinning disc of gas and dust started to form into clumps. Closer to the Sun these materials included rock and iron, but beyond the frost line, which lies between the orbits of Mars and Jupiter, there were solid 'ices' like water, methane and ammonia.

3. Clumps collide and merge
Clumps combined through collisions and started to become the building blocks of planets. Over millions of years these planetesimals increased in size through more collisions. The four giant planets beyond the frost line grew big enough to amass hydrogen and helium.

4. The planets take shape
Solar wind from the Sun dispersed any remaining gas from the Solar System, and planet formation was almost done. Uranus and Neptune are thought to have formed after the dust was swept away from the inner Solar System to its outer regions.

5. Planets get in formation
Uranus and Neptune might have formed farther in between Jupiter and Saturn and later migrated out to their final positions today over hundreds of millions of years. Uranus was likely hit by an impactor when its moons and rings were still forming.

63

UNDERSTANDING THE SOLAR SYSTEM

This false-colour image by Hubble shows bright clouds in Neptune's atmosphere in orange

solid hybrid. No superionic matter has ever actually been observed, but it's thought to exist inside giant planets. In their research paper the team say that ammonia hemihydrate will likely precipitate out of ammonia-water mixtures at high enough pressures. The reason this study is particularly important is because this result emerged from modelling a mixture of ices, instead of individual compounds, which has been the case for studies in the past.

But what else did the team's study show? "We calculated that ammonia hemihydrate has a lower density than pure water ice at the same pressure. It would then form a well-defined layer above an icy sea," explains Hermann. Imagine a solid-liquid ammonia layer above a slushy, frozen ocean at 3 million Earth atmospheres. However, he also says that truly understanding what layers would actually form, if any, would also require adding methane and excess hydrogen to the simulations. In the meantime, the team is working on simulations of superionic ammonia hemihydrate.

Though physically recreating the interiors of the ice giants may be challenging, it's not completely impossible, as has been discovered. An international team led by Dr Dominik Kraus of the University of Rostock and Helmholtz-Zentrum Dresden-Rossendorf fired a powerful X-ray laser at pieces of polystyrene at the SLAC National Accelerator Laboratory. The polystyrene was meant to be a stand-in for methane inside ice-giant mantles – both made of carbon and hydrogen – and the laser created two shock waves within it. Under those conditions, the shock waves overlapped, creating pressures of 1.5 million atmospheres and temperatures

> "Physically recreating the interiors of the ice giants is challenging"

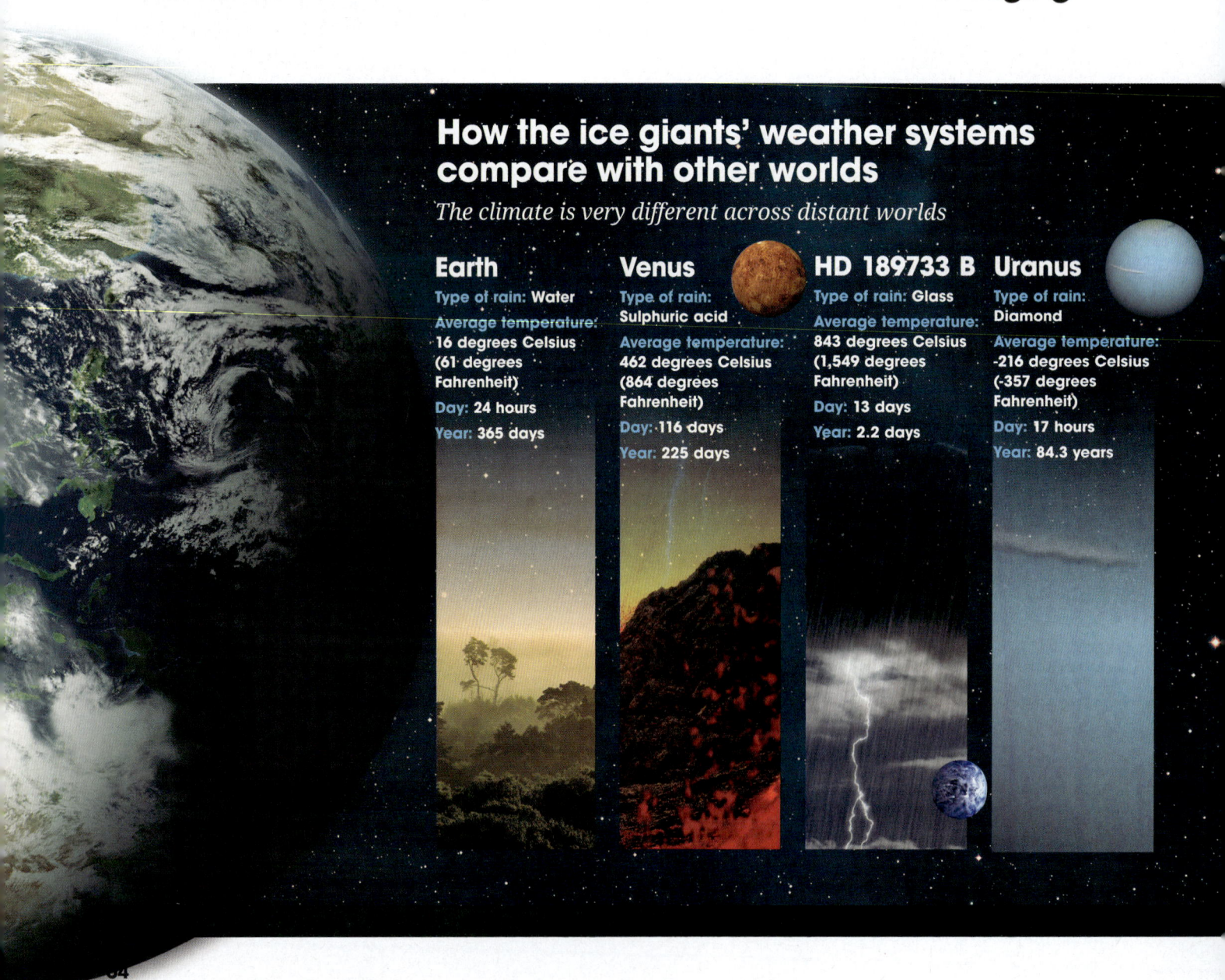

How the ice giants' weather systems compare with other worlds
The climate is very different across distant worlds

Earth
Type of rain: Water
Average temperature: 16 degrees Celsius (61 degrees Fahrenheit)
Day: 24 hours
Year: 365 days

Venus
Type of rain: Sulphuric acid
Average temperature: 462 degrees Celsius (864 degrees Fahrenheit)
Day: 116 days
Year: 225 days

HD 189733 B
Type of rain: Glass
Average temperature: 843 degrees Celsius (1,549 degrees Fahrenheit)
Day: 13 days
Year: 2.2 days

Uranus
Type of rain: Diamond
Average temperature: -216 degrees Celsius (-357 degrees Fahrenheit)
Day: 17 hours
Year: 84.3 years

SECRETS OF
THE ICE GIANTS

Inside Neptune

1 Hydrogen-helium atmosphere
It has a hydrogen-helium atmosphere, albeit with a more dynamic climate and prominent upper clouds. The nature of the deep-blue colour is unknown.

2 Silicate iron-nickel core
Neptune has a rocky iron-nickel core that's at least as massive as Earth. Pressures at its centre could reach 7 million Earth atmospheres.

3 Fluid icy mantle
Neptune's bulk mass is composed of a mantle made of ammonia, water, methane and other ices. Recent experiments suggest a carbon ocean.

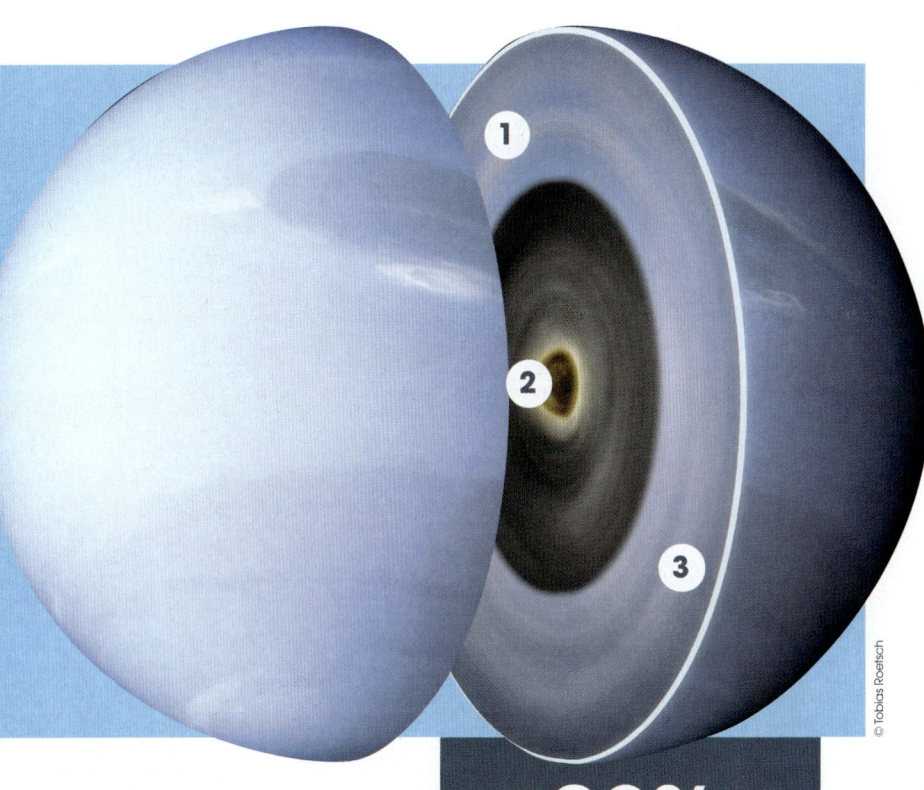

90% Hydrogen
19% Helium
1% Methane

Neptune
Type of rain: Diamond
Average temperature: -214 degrees Celsius (-353 degrees Fahrenheit)
Day: 16 hours
Year: 164.8 years

OGLE-TR-56B
Type of rain: Iron
Average temperature: 1,699 degrees Celsius (3,090 degrees Fahrenheit)
Day: Unknown
Year: 1.2 days

Titan
Type of rain: Methane
Average temperature: -179 degrees Celsius (-290 degrees Fahrenheit)
Day: 16 days
Year: 29 years

of 5,000 degrees Kelvin for fractions of a second – briefly mimicking the conditions inside an ice-giant mantle. The team was surprised to discover that diamond was created, albeit nanometres in size. They theorise that in the more sustained conditions of an ice-giant mantle – around 10,000 kilometres (6,214 miles) down – the diamonds will grow to a larger size as the methane breaks down into hydrogen and carbon and precipitate down to the core. Previous teams have used methane inside laboratory diamond anvil cells to create diamond, but under lower temperatures and pressures. However, the results were always inconclusive.

In another study, Lawrence Livermore National Laboratory scientists subjected a diamond to 1.1 million Kelvin and 40 million atmospheres to recreate the conditions inside giant planets. Results

UNDERSTANDING THE SOLAR SYSTEM

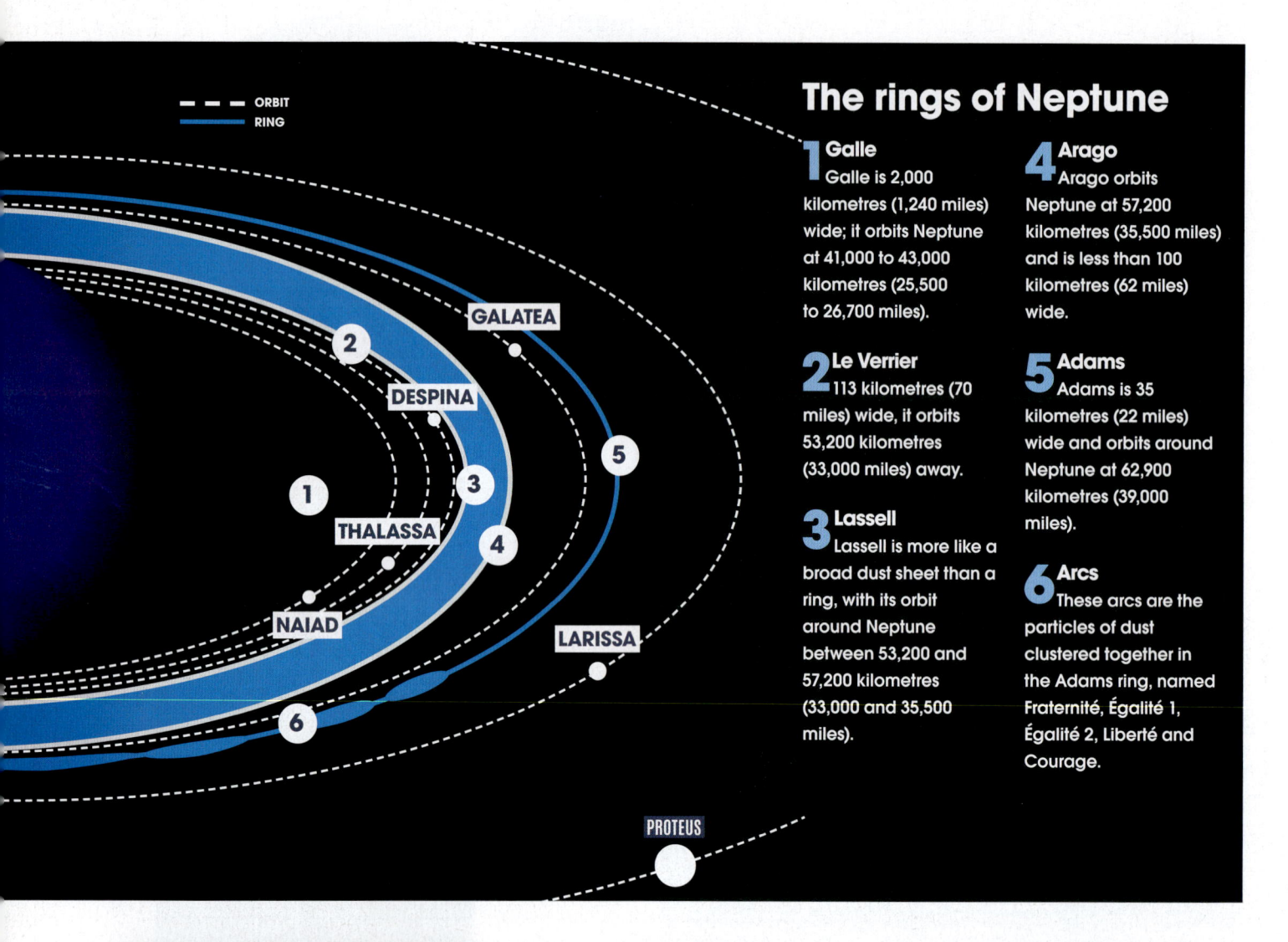

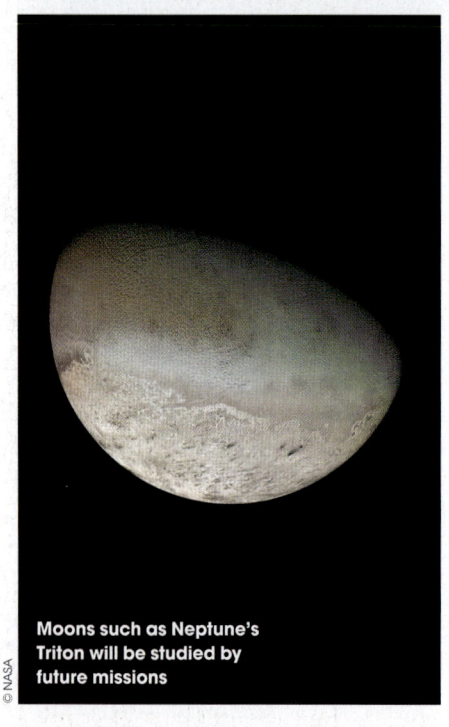

The rings of Neptune

1 Galle
Galle is 2,000 kilometres (1,240 miles) wide; it orbits Neptune at 41,000 to 43,000 kilometres (25,500 to 26,700 miles).

2 Le Verrier
113 kilometres (70 miles) wide, it orbits 53,200 kilometres (33,000 miles) away.

3 Lassell
Lassell is more like a broad dust sheet than a ring, with its orbit around Neptune between 53,200 and 57,200 kilometres (33,000 and 35,500 miles).

4 Arago
Arago orbits Neptune at 57,200 kilometres (35,500 miles) and is less than 100 kilometres (62 miles) wide.

5 Adams
Adams is 35 kilometres (22 miles) wide and orbits around Neptune at 62,900 kilometres (39,000 miles).

6 Arcs
These arcs are the particles of dust clustered together in the Adams ring, named Fraternité, Égalité 1, Égalité 2, Liberté and Courage.

Moons such as Neptune's Triton will be studied by future missions

suggested that at the bottom of the ice giants' mantles could lie a liquid-carbon layer with chunks of floating diamond.

Kraus' team are now working on follow-up experiments. "Our efforts have now turned to looking at what happens when we reduce the carbon concentration in our samples and add other light elements that are also present inside Neptune and Uranus, such as oxygen or nitrogen," he says. They're also figuring out ways to safely capture the nanodiamond particles, which travel at incredibly high speeds and are currently only detected via spectroscopy.

The work done so far may help solve another mystery: why Uranus radiates less excess heat than it receives from the Sun compared to all the other giant planets. Even Neptune, which is farther from the Sun, radiates 2.6 times as much heat as it receives. This heat energy may be what drives Neptune's storm systems and gives it a more dynamic climate than Uranus. It's been suggested that somewhere inside Uranus is a thermal boundary layer that stops heat from escaping. Hermann says that if the planet's layers are more stratified than thought, "it's not inconceivable that ammonia hemihydrate, or similar strongly bound ionic phases, could form such a layer." Such work could ultimately help in understanding not only the ice giants, but all the giant planets. As Fortney says: "The deep cores of Jupiter and Saturn may actually resemble Uranus and Neptune, but at much higher pressures and temperatures."

Kulvinder Singh Chadha
Space science writer
Kulvinder is a freelance science writer, outreach worker and former assistant editor of Astronomy Now. He holds a degree in astrophysics.

SECRETS OF
THE ICE GIANTS

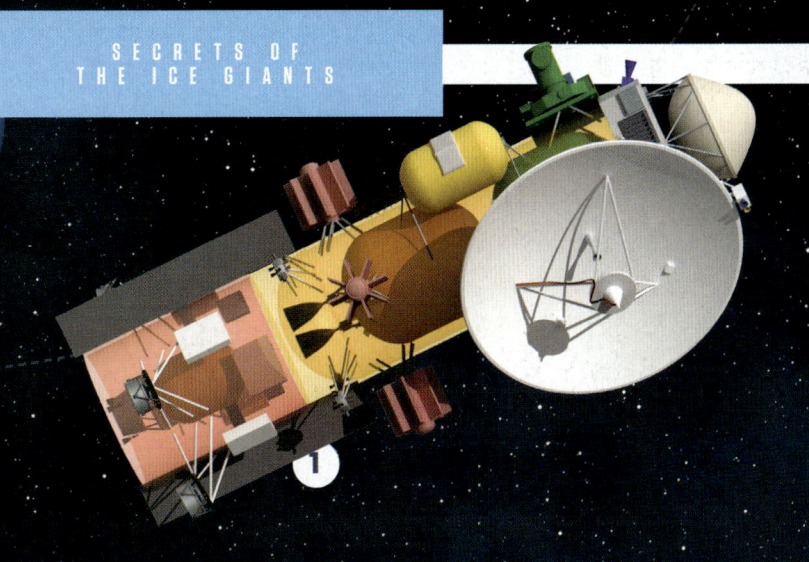

Missions to Uranus and Neptune

NASA reports suggest several possible future spacecraft to the ice giants

Neptune orbiter with probe
Launch date: 2030s
Mission length: 15 years

1 A minimum 50-kilogram science payload would be required for a mission to Neptune. This would include both the orbiter and probe's science packages. The orbiter would study Neptune's moons – particularly Triton, which is a captured Kuiper Belt object. It would also study Neptune's weather systems, magnetosphere and solar wind particles at this distance. The atmospheric probe would be required to accurately measure the abundance of noble gases, including hydrogen and helium, and other elements.

Uranus orbiter only
Launch date: 2030s
Mission length: 15 years

2 This would weigh three times as much as the other concepts, but carry five times as many instruments. Alongside a narrow-angle camera and instruments for measuring magnetic fields and interior atmospheric structure, the orbiter will have spectrometers, a dust detector and other devices. It would have a wide-angle camera for snapping wide vistas of the planet, its moons and rings. It may clear up long-standing mysteries. One big mystery of Uranus is why it radiates so little heat compared to the other giant planets.

Uranus orbiter with probe
Launch date: 2030s
Mission length: 15 years

3 An orbiter and probe would have a major advantage over a flyby. While the probe studied the planet's atmospheric layers, the orbiter could study the moons and ring system in great detail. As a comparison to Cassini, unexpected discoveries such as plumes of water ice on Enceladus were a fortunate consequence. A Uranus orbiter-probe would be almost the same as the Neptune orbiter-probe concept. It's thought both planets' moons harbour water in some form. This would be of value in understanding both planets' environments.

Uranus flyby with probe
Launch date: 2030s
Mission length: 10 years

4 A flyby, weight for weight, would be the cheapest ice-giant mission option, and it would achieve many of the same science objectives. However, a flyby mission wouldn't have much time to study the planet, its moons or ring system. It would have to work fast and release its probe. As with Neptune, the probe would be required to measure the abundance of hydrogen, helium, heavier noble gases and other elements such as volatiles in the atmosphere. Although not the same as Neptune, a lot could be inferred from it by studying Uranus.

67

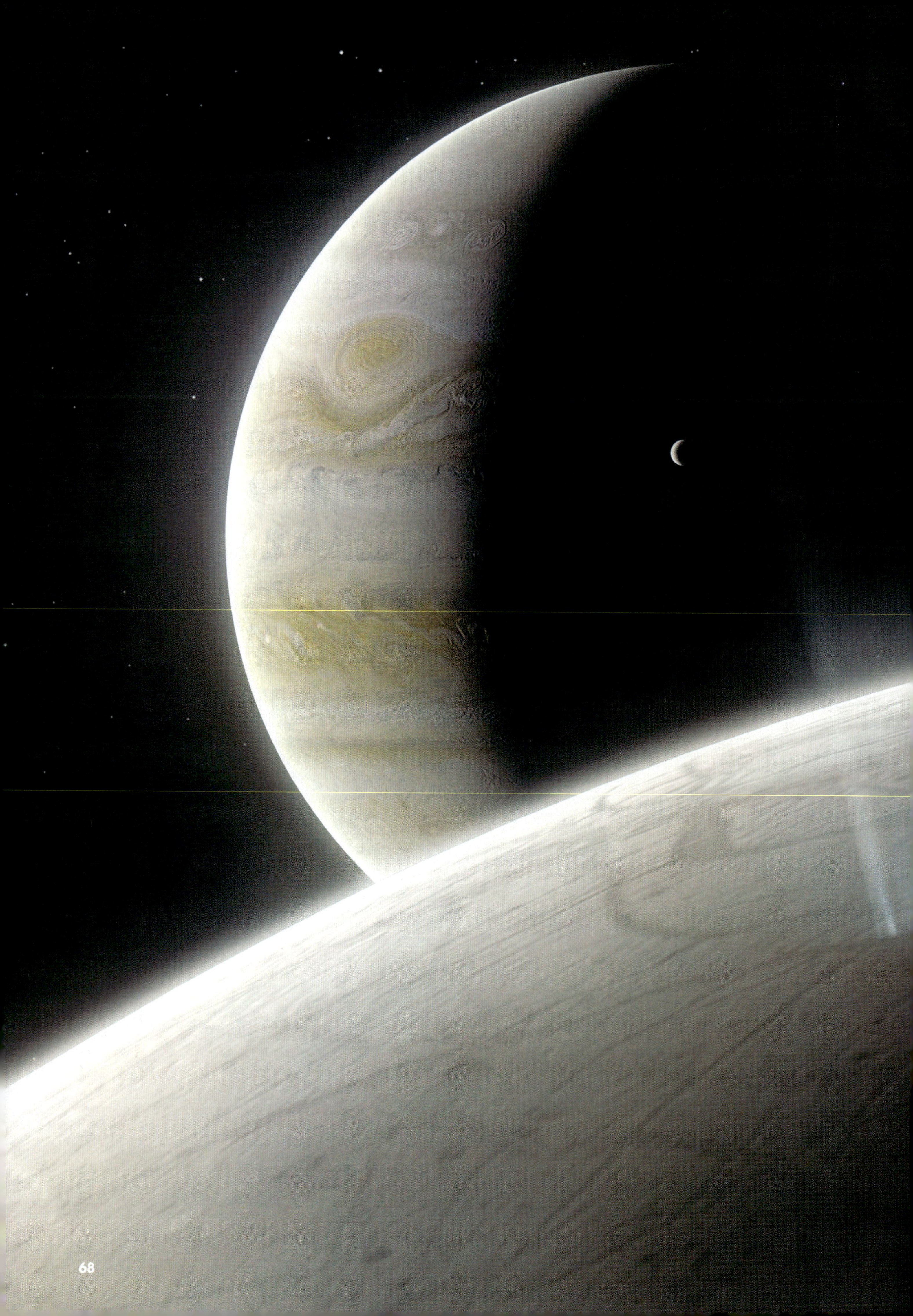

Moons of the
Solar System

UNDERSTANDING THE SOLAR SYSTEM

Complete guide to the Moon

Even though we know more about our natural satellite than any other celestial body – and have even visited it – the Moon continues to fascinate us

COMPLETE GUIDE TO THE MOON

Because we can easily discern features on the Moon with the naked eye, it's been a source of wonder to us since ancient times. The Moon is the brightest object in our sky after the Sun, and influences everything from our oceans to our calendars. It's always been 'the Moon' because we didn't know that there were any others. Once Galileo discovered in 1610 that Jupiter had satellites, we've used the word 'moon' to describe celestial bodies that orbit larger bodies, which orbit stars. Since the Moon has always been so present it might not seem worth studying, yet there's a reason why we continue to return to it – we still have plenty to learn from our satellite.

The Moon is the fifth-largest and second-densest satellite in the Solar System. Its diameter is 27 per cent of Earth's at 3,476 kilometres (2,160 miles), while its mean density is 60 per cent that of Earth's. This makes the Moon the largest satellite in size relative to the planet that it orbits. The Moon is also unusual because its orbit is more closely aligned to the plane of the ecliptic – the plane in which Earth orbits.

Most planetary satellites orbit closer to their planet's equatorial plane, but the Moon is inclined from the plane of the ecliptic by approximately 5.1 degrees.

Its average distance from Earth is 384,400 kilometres (239,000 miles), and it completes an orbit once every 27.3 days. The Moon is in synchronous rotation with Earth – its rotation and orbital period are the same – so the same side is almost always facing our planet. This is called the 'near side' of the Moon, while the opposite side is the 'far' or 'dark' side, although it gets illuminated just as often as the near side. This hasn't always been the case: the Moon used to rotate faster but the influence of Earth caused it to slow and become locked.

Although we say that we can only see one side of the Moon at a time, that's not strictly true. The Moon's orbit isn't

> "The Moon is the fifth largest and second-densest satellite in the Solar System"

The Moon and tides

Along with the Sun, the Moon exerts serious force on Earth's tides. Whether the tides vary widely or not much at all has a lot to do with the interactions between the solar and lunar cycles. When they are together, their combined effects produce tidal variations called spring tides – high tides are very high and low tides are very low. If the Sun and Moon are on the opposite sides of the sky, they nullify each others' effects, producing neap tides – with little variation.

1 Spring tide
During new and full Moon, both the Sun and Moon exert a strong effect, producing spring tides.

2 Neap tide
On first and third quarter Moon, the Sun and Moon have little effect on tidal range, leading to neap tides.

UNDERSTANDING THE SOLAR SYSTEM

Measuring up the Moon by its diameter

Ganymede
Jupiter
5,268 kilometres
(3,300 miles)

Titan
Saturn
5,150 kilometres
(3,200 miles)

Callisto
Jupiter
4,821 kilometres
(3,000 miles)

Io
Jupiter
3,642 kilometres
(2,300 miles)

The Moon
Earth
3,476 kilometres
(2,160 miles)

Europa
Jupiter
3,122 kilometres
(1,940 miles)

Triton
Neptune
2,700 kilometres
(1,700 miles)

Titania
Uranus
1,578 kilometres
(980 miles)

© NASA

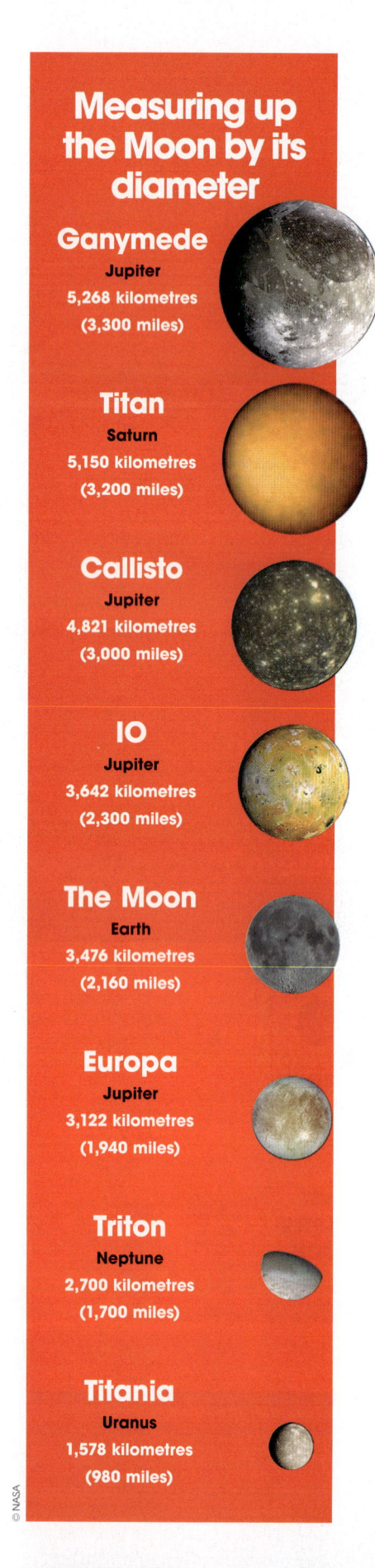

quite circular – it has an eccentricity of 0.0549. The Moon also wobbles a bit along its orbit. These two factors cause some variations in how much of the Moon that we see, called librations. When the Moon is closest to Earth, called the perigee, it orbits slightly slower than it rotates. This means that we can actually get a glimpse of about eight degrees of longitude of its eastern far side. When the Moon is at its furthest point away in its orbit, the apogee, its orbit is slightly faster than its rotation. So we get a glimpse of eight degrees of longitude on its western far side. This is called longitudinal libration.

The Moon also appears to rotate towards and away from Earth. This is due to the 5.1-degree inclination of its orbit, as well as the 1.5-degree tilt of the Moon's equator to the plane of the ecliptic. This results in us seeing about 6.5 degrees of latitude on the far side along both the sides of the poles. The Moon's orbit also means that it appears to move about 13 degrees across the sky each day, and this movement accounts for the lunar phases.

The Moon's gravitational pull has a strong effect on Earth. The most obvious effect to us is the tides. High tide occurs when the level of water at the coastline rises, and low tide occurs when the water rushes further out. While some coastlines experience one high tide and one low tide per day, of equal strength, others have different strengths, timing and numbers of tides. Measuring and predicting these tides is vital for everything from fishing to navigating intercoastal waterways. We use the term 'tides' to describe oceanic tides, but tides also occur on a smaller level in lakes as well as Earth's atmosphere and crust.

Scientists believe that the Moon formed when a huge celestial body about the size of Mars – which has been given the name Theia – impacted with a young Earth. This is known as the giant impact hypothesis. This force sent debris out into Earth's orbit, which fused to form the Moon. However, in 2012 an analysis of rock samples taken from the Moon during the Apollo missions showed that the Moon's composition is almost identical to Earth's. This calls the giant impact hypothesis into question because previously we thought that at least some of the Moon's material had to have come from Theia.

Two sides of the Moon

As opposed to the near side, the far side is covered with craters and very little maria. This may be because it's hotter on that side, or because it experienced a collision.

The near side is mostly covered in dark areas that were originally thought to be seas, called maria. The lighter areas are called the lunar highlands.

COMPLETE GUIDE TO THE MOON

How the Moon was made

1 Theia nears Earth
A Mars-sized object on an unalterable collision course with early Earth.

2 Earth is hit
The impactor hit Earth in a head-on collision, vaporising both Theia and the mantle of Earth.

3 Material is thrown out
The vaporised material from both bodies mixed and was thrown outwards.

4 Debris gathers
Smaller objects began to condense out of the vapour while continuing to orbit around Earth.

5 The Moon takes shape
Many of the smaller objects stuck together to form a protomoon in orbit around Earth.

6 Our companion is formed
Eventually all the pieces came together to form the basis of the Moon we see today.

The Moon's orbit

1 First quarter
Half of the Moon is visible in the afternoon and early evening.

2 Waxing crescent
Up to 49 per cent of the Moon is visible in the afternoon and after dusk.

3 New Moon
The first visible crescent in the southern hemisphere, seen after sunset.

4 Waning crescent
Up to 49 per cent is visible just before dawn and in the morning.

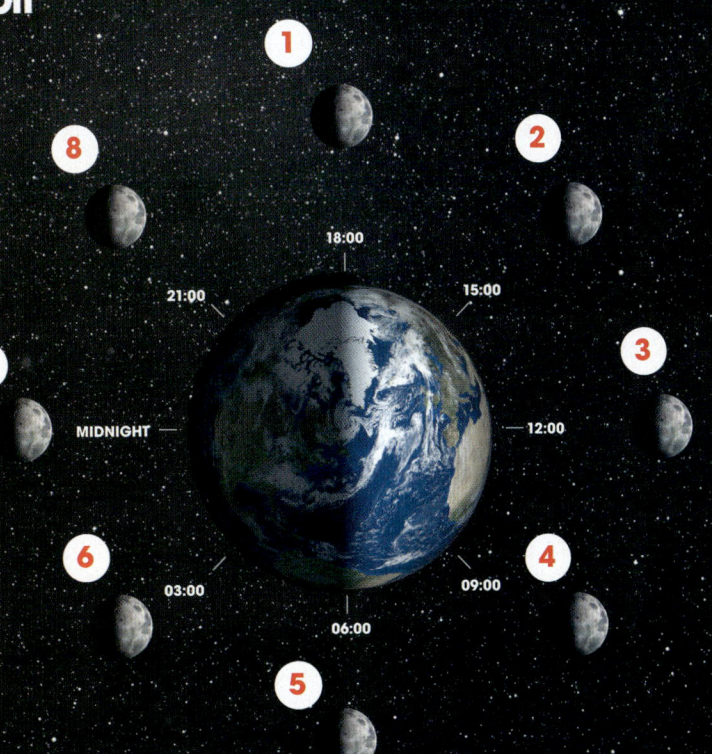

5 Third quarter
Half of the Moon is visible in the late evening and morning.

6 Waning gibbous
Between 51 and 99 per cent of the Moon is visible for most of the evening and in the early morning.

7 Full Moon
The entire Moon is visible all night long.

8 Waxing gibbous
Between 51 and 99 per cent is visible in the later afternoon and most of the evening.

73

UNDERSTANDING THE SOLAR SYSTEM

Inside and out

Earth's natural satellite shares some remarkable similarities with our home planet

When astronauts explored the Moon they discovered it has moonquakes

Although the Moon may seem like a solid rock, it's actually differentiated like Earth; it has a core, a mantle and a crust. The Moon's structure likely came from the fractional crystallisation of a magma ocean that once covered it. This probably happened not long after the Moon was formed, about 4.5 billion years ago. As the magma ocean cooled, its composition changed as the different minerals within the melt crystallised into solids. The denser materials sank, forming the mantle, while less dense materials floated on top and formed the crust.

The core is probably very small, with a radius about 20 per cent the total size of the Moon. By contrast, most differentiated celestial bodies have cores about 50 per cent of their total size. The core itself comprises a solid innermost core that is rich in iron as well as nickel and sulphur, with a radius of 240 kilometres (150 miles). This is surrounded by a fluid outer core with about a 300-kilometre (186-mile) radius. Between the core and the mantle, there's a boundary layer of partially melted iron that has about a 500-kilometre (300-mile) radius. It is also known as the lower mantle. The upper mantle is mafic – rich in magnesium and iron, topped by a crust of igneous rock called anorthosite. It mainly includes aluminium, calcium iron, magnesium and oxygen, with traces of other minerals. We estimate the crust is around 50 kilometres (31 miles) thick.

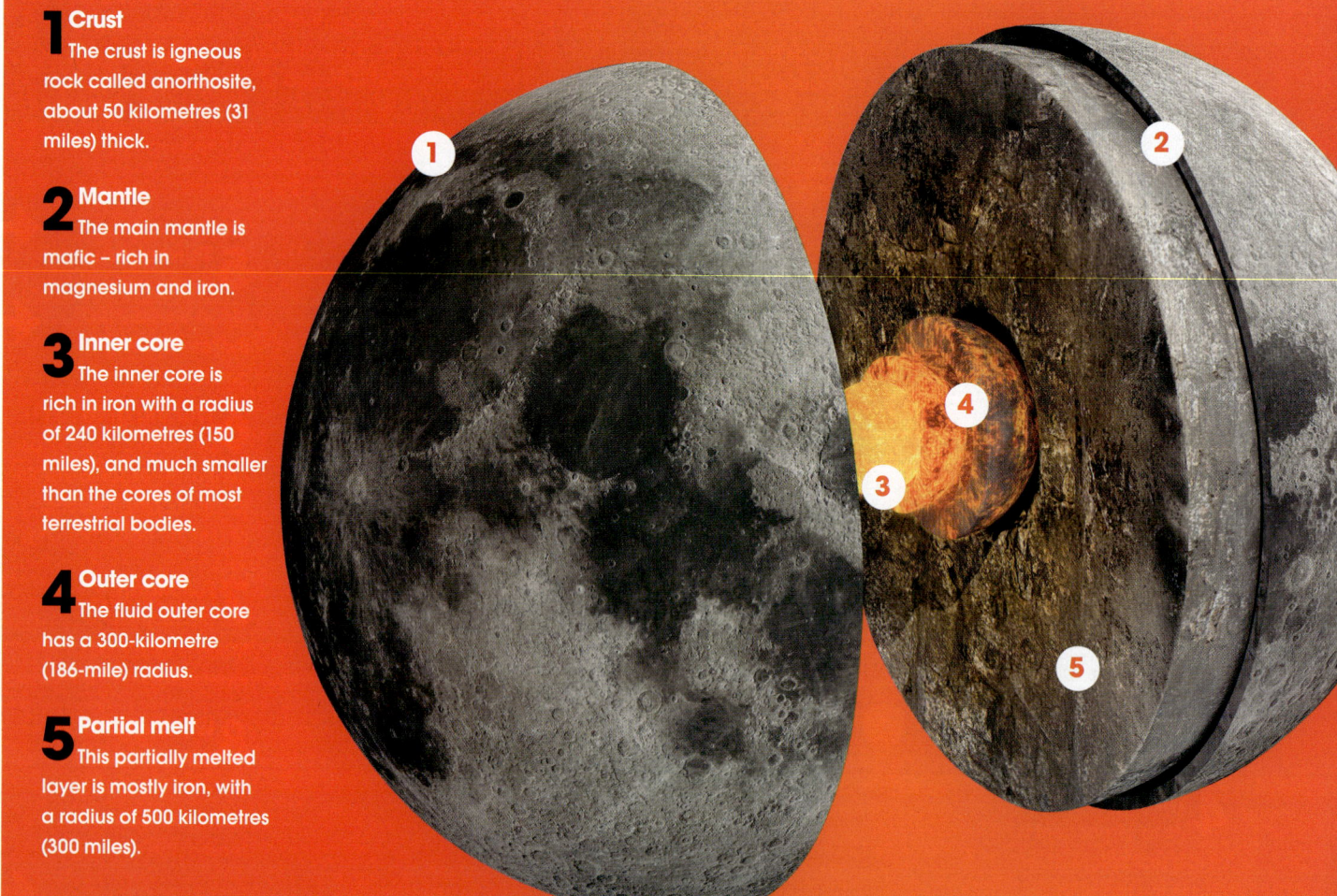

1 Crust
The crust is igneous rock called anorthosite, about 50 kilometres (31 miles) thick.

2 Mantle
The main mantle is mafic – rich in magnesium and iron.

3 Inner core
The inner core is rich in iron with a radius of 240 kilometres (150 miles), and much smaller than the cores of most terrestrial bodies.

4 Outer core
The fluid outer core has a 300-kilometre (186-mile) radius.

5 Partial melt
This partially melted layer is mostly iron, with a radius of 500 kilometres (300 miles).

COMPLETE GUIDE TO THE MOON

The Moon by numbers

400
How many times bigger the Sun is than the Moon. It's also about 400 times further away from Earth, which is why they look the same size in the sky

29.5 days
The length of a lunar month, longer than the amount of time it takes the Moon to orbit Earth because Earth is moving, too

12
The number of people who have set foot on the Moon

3.8 centimetres
The distance the Moon moves away from Earth each year

13 hours
The amount of time it takes to reach the Moon by rocket

16.6 kilograms
The amount you would weigh on the Moon if you weighed 100 kilograms on Earth

The magnetic field mystery

The Moon has an external magnetic field – it's less than one-hundredth that of Earth's magnetic field. It's not a dipolar magnetic field like Earth, which has a field that radiates from the north and south poles. Researchers believe the Moon once had a dipole magnetic field, created by a dynamo – a convecting liquid core of molten metal. But we aren't sure what powered that dynamo. It could have worked like Earth's dynamo. Earth's dynamo powers itself as elemental radioactive decay maintains convection in the core. The Moon could also have had a dynamo powered by the cooling of elements at the core.

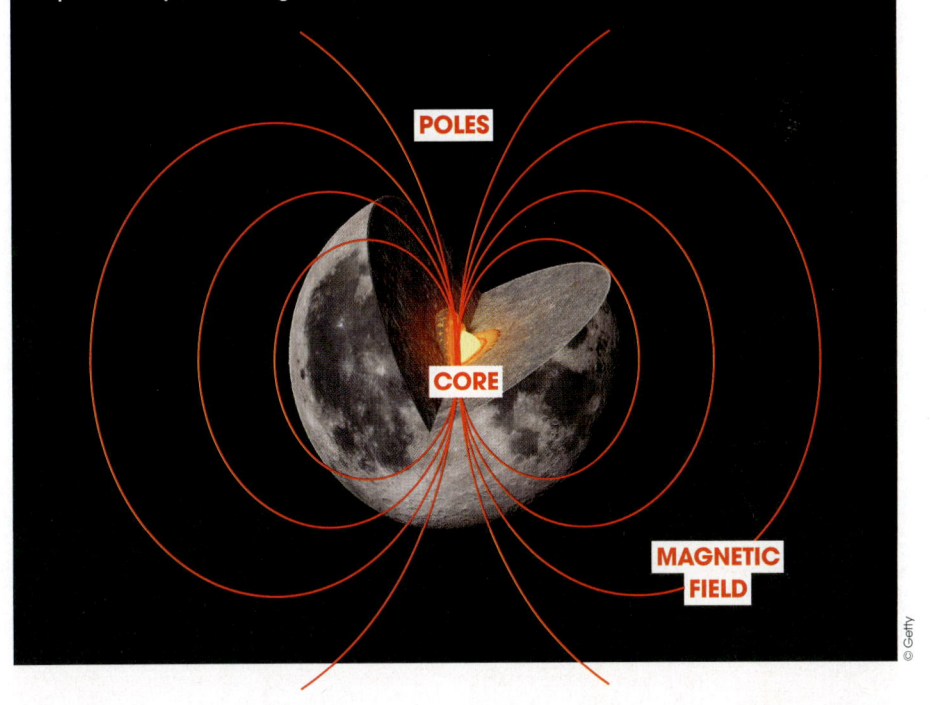

POLES · **CORE** · **MAGNETIC FIELD**

© Getty

The Moon has no plate tectonics, but it does have seismic activity. When astronauts visited the Moon they discovered that there are moonquakes – the Moon's equivalent of earthquakes. Moonquakes aren't as strong as earthquakes, but they can last longer because there's no water to lessen the effects of the vibrations. Seismometers showed that the strongest moonquakes are about 5.5 on the Richter scale. There are four different types of moonquakes: shallow, deep, thermal and meteorite. Shallow ones occur 20 kilometres (12 miles) below the surface, while deep moonquakes can be as deep as 700 kilometres (435 miles). These deep moonquakes are probably related to stresses on the Moon caused by its eccentric orbit and gravitational interactions between it and Earth. Thermal earthquakes occur when the Moon's crust heats and expands. Shallow moonquakes are the strongest and most common. Nearly 30 were recorded between 1972 and 1977 by seismometers on the surface. This data has helped us to determine the Moon's internal composition.

The dominating feature on the near side of the Moon's surface, called maria, are the result of ancient volcanic activity. These vast, dark plains are basalts – igneous rock that formed after lava erupted due to partial melting within the mantle. These basalts show that the Moon's mantle is much higher in iron than Earth's, and has a varied composition. Some basalts are very high in titanium, while others are higher in minerals like olivine.

These basalt maria have influenced the Moon's gravitational field because they're so rich in iron. The gravitational field contains mascons, positive gravitational anomalies that influence how spacecraft orbit the Moon. The maria can't explain all of the mascons that have been tracked by the Doppler effect on the radio signals emitted by spacecraft that orbit the Moon. And there are also some large maria without associated mascons.

On the surface

The surface of the Moon is about contrasts: light and dark, hot and cold

The landscape of the Moon is dominated by three main features: maria, terrae and craters. The basalt maria appear dark due to their high iron content and are much more prevalent on the near side of the Moon. Other volcanic features on the surface include domes and rilles. Domes are shield volcanoes that are round and wide with gentle slopes, while rilles are twisting sinuous formations caused by channels of flowing lava.

The lighter areas on the Moon are called terrae, or lunar highlands. They are made up of anorthosite, the type of igneous rock that dominates the overall crust of the Moon. While this type of rock can be located in some places on Earth, it's not generally found on the surface due to plate tectonics and deposits. These highlands reflect light from the Sun and make it appear that the Moon is glowing at night.

Both the maria and terrae have impact craters which were formed when asteroids and comets struck the surface of the Moon. These craters range in size from very tiny to massive. It is estimated that there are around 300,000 craters on the near side of the Moon that are wider than one kilometre (0.62 miles). The largest impact crater, called the South Pole-Aitken Basin, is about 2,500 kilometres (1,550 miles) in diameter and 13 kilometres (eight miles) deep. The biggest craters also tend to be the oldest, and many are covered in smaller craters. Younger craters have more sharply defined edges, while older ones are often softer and rounder. If the impact was especially large, material may be ejected from the surface to form secondary craters.

In some cases, the basalt eruptions flowed into or over large impact craters called basins. In general, the terrae have far more craters because the maria are younger in age than the terrae. While the Moon isn't much younger than Earth, our planet has processes that continue to change its surface over time, like erosion and plate tectonics. The Moon doesn't experience these, which is why some impact craters are up to 500 million years older than the basalt filling them.

The loose soil on the Moon is called regolith. It's powdery and filled with small rocks. Over time, impacts from meteors, as well as space weathering (solar wind, cosmic rays, meteorite bombardment and other processes), break down the rocks and grind them into dust. Aside from the basalt and anorthosite rocks, there are also impact breccias – rock fragments that were welded together by meteor impacts – and glass globules from volcanic activity.

Although you may sometimes see the term 'lunar atmosphere', the Moon is actually considered to exist in a vacuum. There are particles suspended above the surface, but the density of the Moon's atmosphere is less than one hundred trillionth that of Earth's atmosphere. What little atmosphere there is gets quickly lost to outer space, and is constantly replenished. Two processes help to replenish the Moon's atmosphere: sputtering and outgassing. Sputtering occurs when sunlight, solar wind and meteors bombard the surface and eject particles. Outgassing comes from the radioactive decay of minerals in the crust and mantles, which can release gases like radon.

The Moon has a very minor axial tilt, so there aren't seasons in the same way that we have them here on Earth. However, temperatures on the Moon can change dramatically because there's no atmosphere to trap heat, and portions of the Moon may be either in full sunlight or total darkness depending on where it

Exploring the Moon: The Past

Apollo 11
21 July 1969
NASA astronauts Buzz Aldrin and Neil Armstrong became the first humans to set foot on another body in space when they landed on the Moon in 1969.

Apollo 12
19 November 1969
The second spacecraft to land on the Moon, Apollo 12, used a Doppler effect radar technique to land within walking distance of the Surveyor 3 probe.

Apollo 14
5 February 1971
Apollo 14's commander was Alan Shepard who, a decade earlier on 5 May 1961, had become the second person in space after Yuri Gagarin and the first American.

Apollo 15
30 July 1971
NASA deemed this landing the most successful so far out of its manned missions. It is also the first of the longer missions to the Moon, called 'J missions', staying for three days.

Apollo 16
21 April 1972
Apollo 16 was the first spacecraft to land in the highlands on the Moon, which let the astronauts gather older lunar rocks.

Apollo 17
11 December 1972
This last manned mission to the Moon carried the Traverse Gravimeter Experiment, which measured relative gravity at different sites on the Moon.

Luna 1
4 January 1959
This Soviet probe was the first to reach the vicinity of the Moon and the first to break out of geocentric orbit. But it didn't impact the Moon as had originally been planned.

Luna 21
15 January 1973
This Soviet spacecraft landed on the Moon and carried a lunar rover, Lunokhod 2. It performed numerous experiments and sent back more than 80,000 images.

Luna 24
22 August 1976
This was the last of the Luna missions, landing near Mare Crisium to recover samples. It was the last spacecraft to have a soft landing on the Moon until 2013.

COMPLETE GUIDE TO THE MOON

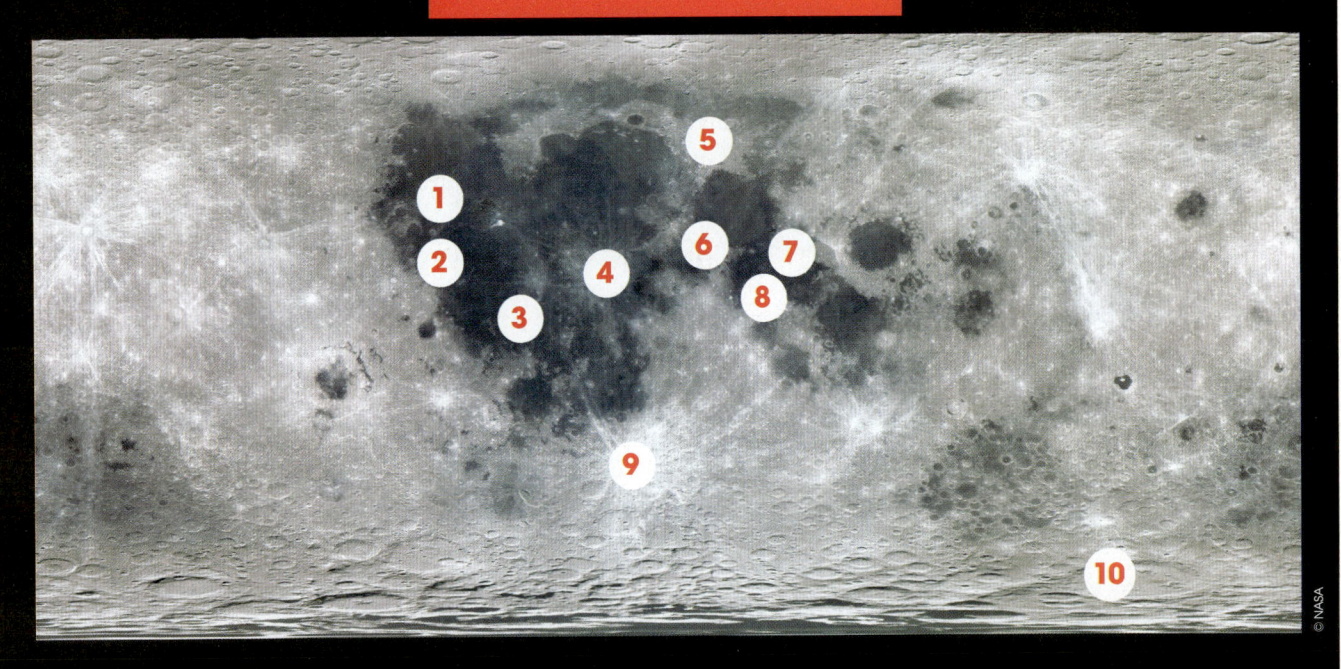

A lunar world tour

1 Oceanus Procellarum
This mare is so large that it was deemed an ocean, covering about 4,000,000 square kilometres (1,500,000 square miles).

2 Luna 9
This site marks the first soft landing of an unmanned spacecraft on the Moon, launched by the Soviet space program on 31 January 1966.

3 Surveyor 1
The first American soft Moon landing happened here, launched on 30 May 1966.

4 Copernicus
This crater is well known because it can be easily seen from Earth. It is a younger crater, about 800 million years old, with a prominent system of ejecta rays.

5 Vallis Alpes
This lunar valley bisects a mountain range called the Montes Alpes, and extends 166 kilometres (103 miles).

6 Montes Apenninus
This mountain range is about 600 kilometres (370 miles) long and has peaks up to 5 kilometres (three miles) high.

7 Mare Tranquillitatis
This mare was the landing site for the Apollo 11 spacecraft. It is slightly bluish because it has a high metal content.

8 Apollo 11
Neil Armstrong and Buzz Aldrin became the first men to set foot on the Moon, on 21 July 1969 as part of NASA's Apollo program.

9 Tycho
This distinctive crater has ejecta rays visible from Earth during a full Moon, reaching more than 1,000 kilometres (621 miles) from the crater.

10 Schrödinger
This huge crater near the south pole can only be viewed from orbit. It is 312 kilometres (194 miles) in diameter.

is in its rotation. Full sunlight can mean temperatures of greater than 100 degrees Celsius (212 degrees Fahrenheit). But at the end of the lunar day the temperature can drop by hundreds of degrees. There are also big differences in temperatures depending on the surface features. For example, the Moon is coldest in its deepest craters, which always remain in darkness. The coldest temperature ever recorded in the Solar System by a spacecraft was measured by the Lunar Reconnaissance Orbiter in the Hermite Crater near the Moon's north pole at -248 degrees Celsius (-414 degrees Fahrenheit).

The last manned mission to the Moon took place in December 1972

UNDERSTANDING THE SOLAR SYSTEM

"Analysis of rock samples taken from the Moon during the Apollo missions showed that the Moon's composition is almost identical to Earth's"

American boots on the Moon ended the Space Race

COMPLETE GUIDE TO THE MOON

The remains of Soviet probe Luna 24 rest in Mare Crisium

UNDERSTANDING THE SOLAR SYSTEM

Humans will return to the Moon in 2025 and explore sites for a permanent base

Exploring the Moon: Present and future

We've been studying the Moon for over 50 years, and thanks to a host of pioneering missions we now know more about our satellite than ever before

Although there hasn't been a manned mission to the Moon since 1972 and there were no soft landings at all until 1966, we're still exploring our satellite. Currently the Lunar Reconnaissance Orbiter (LRO) is still circling the Moon. It launched on 18 June 2009, the first NASA mission to the Moon in more than a decade. The LRO is meant to be a precursor to future manned missions, and was originally designed to spend just a year in orbit. However, the mission was extended several times. It was designed to extensively map the Moon in high resolution, explore the potential of ice in the polar regions, study the deep space radiation, and continue to map the surface of the Moon. The other current NASA mission is ARTEMIS, an extension of an earlier satellite mission. Two small probes have been orbiting the Moon together since summer 2011, having previously performed lunar and Earth flybys.

The Lunar Crater Observation and Sensing Satellite (LCROSS) was launched along with the LRO and considered an inexpensive way to look for water ice, and it was successful. The LCROSS discovered ice in the Cabeus crater near the Moon's south pole after its upper stage impacted as planned on 9 October 2009. Two small spacecraft under the name GRAIL A and GRAIL B were launched on 10 September 2011 and impacted on 17 December 2012, having collected data to help understand how terrestrial planets have evolved. Japan, India and China have all had lunar probes in the last six years as well.

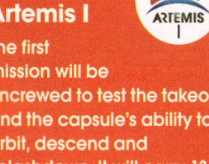

Main mission objectives

Long-term presence
Following Apollo 17's three-day presence on the Moon, Artemis will send astronauts there for weeks.

Equality
A female astronaut hasn't set foot on the Moon yet. This mission will demonstrate the increasing role women have played in space missions since the Apollo era.

Partnerships
NASA has collaborated with private companies such as SpaceX and Boeing. These show space travel's shift towards commercialisation.

Technology
NASA is always learning from past missions; the spacecraft and spacesuits have been tailored to the Moon mission, exhibiting the latest in space technology.

Knowledge
Collecting further information about the lunar surface and deep space, NASA hopes to become better prepared for later missions back to the Moon and further afield.

Resources
Access to the lunar surface provides the opportunity to search for rare minerals and exploit resources. Hydrogen and oxygen could be used as rocket fuel to travel from the Moon.

2022 Artemis I
The first mission will be uncrewed to test the takeoff and the capsule's ability to orbit, descend and splashdown. It will carry 13 small satellites to perform experiments and technology demonstrations. The craft will orbit the Moon for six days, collecting performance data.

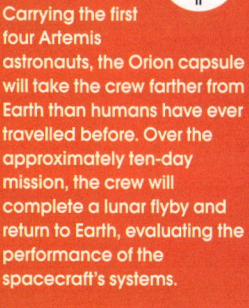

2024 Artemis II
Carrying the first four Artemis astronauts, the Orion capsule will take the crew farther from Earth than humans have ever travelled before. Over the approximately ten-day mission, the crew will complete a lunar flyby and return to Earth, evaluating the performance of the spacecraft's systems.

2025 Artemis III
This will see the next man and first woman step onto the lunar surface. Providing previous missions have been successful, the astronauts will shoot towards the Moon, using the lunar lander to lower two people to the south polar region. They will remain on the Moon for around a week.

80

COMPLETE GUIDE TO THE MOON

Earth to the Moon
Jump on board the Orion as we follow the route planned for the Artemis astronauts

1 Launch day
Scheduled to launch in 2025, the third Artemis mission and second with crew on board will launch from the Kennedy Space Center in Florida. It will be monitored by the nearby Launch Control Center.

2 Entering orbit
Once the rocket has taken Orion into orbit, its engines shut down and it will separate from the capsule. These rocket components will fall towards the Pacific Ocean. Orion will then deploy its solar arrays.

3 Trans-lunar injection
Having made it into Earth orbit, the Orion vehicle will head to the Moon. During a 20-minute burn, the engines will fire to increase the speed, displacing the spacecraft from its low-Earth orbit.

4 To deep space
Set on a precise trajectory, Orion will travel over 384,000 kilometres (239 miles). This needs to account for factors such as gravity and the movement of the Moon. Artemis I will have tested the planned path.

5 Lunar flyby
A main engine burn 185 kilometres (115 miles) above the Moon's surface will put Orion on a trajectory to intercept the orbit of the planned Lunar Gateway space station, set to launch in November 2024.

6 Moon landing
Having docked with Gateway, the crew may need to inspect it and collect supplies. While two astronauts will stay aboard the spacecraft in orbit, the other two will transfer to a lander vehicle.

7 Lunar exploration
The astronauts will remain on the Moon for roughly seven days. As an area where water ice is present, they will explore the suitability of the lunar south pole for a permanent Moon base.

8 Ascent
Having carried out experiments on the Moon, the astronauts will reboard the Human Landing System and return to Gateway. Taking samples with them, they will return to Orion for the journey home.

9 Splashdown
After spending less than 30 days in space, the parachuted capsule will return to Earth, splashing down in the Pacific Ocean. NASA will have a team ready to retrieve the crew and the capsule.

"The astronauts will remain on the Moon for roughly seven days"

UNDERSTANDING THE SOLAR SYSTEM

Strangest moons in the Solar System

Some of the most fascinating worlds in our cosmic neighbourhood are not planets, but the moons that orbit around them

All but two of our Solar System's planets have satellites of one sort or another. Earth's Moon, a beautiful but stark, dead world shaped by ancient volcanoes and countless impact craters, is undoubtedly the most familiar, but it's far from being the most interesting.

Each of the outer Solar System's giant planets is accompanied by a large retinue of satellites, many of which formed at the same time and from the same ice-rich material as the planets that host them. Although far from the Sun and starved of heat and light, they nevertheless show as much variety as the planets themselves.

Here All About Space takes a trip to visit some of the strangest and most exciting of these astonishing worlds. Some, such as Jupiter's Callisto and Saturn's Mimas, have been frozen solid for billions of years but bear extraordinary scars from exposure to bombardment from space. Others, such as Saturn's shepherd moons Pan and Atlas and Neptune's lonely Nereid, have been affected throughout their history by interactions with their neighbours.

Most excitingly, some of these exotic worlds have been heated by powerful tidal forces from their parent planets, triggering phases of violent activity like those which shaped Miranda, Uranus' Frankenstein moon. In some cases these forces are still at work today, creating fascinating bodies such as Jupiter's tortured Io and Saturn's icy Enceladus, whose placid exterior may even hide the greatest secret in the Solar System: extraterrestrial life itself.

Giles Sparrow
Space science writer

Giles has degrees in astronomy and science communication and has written books and articles on all aspects of the universe.

STRANGEST MOONS IN THE SOLAR SYSTEM

Enceladus The ring bearer

Mass: 1.1×10^{20}kg (2.4×10^{20}lbs) **Diameter:** 504km (313 miles) **Parent planet:** Saturn
Discovered: 1789, William Herschelz

Since NASA's Cassini probe arrived at Saturn in 2004, the ringed planet's small inner satellite, Enceladus, has become one of the most intensely studied and debated worlds in the entire Solar System. It owes its new-found fame to the discovery of huge plumes of water ice erupting into space along fissures in its southern hemisphere – a sure sign of liquid water lurking just beneath the moon's thin, icy crust.

The strange activity of Enceladus was suspected before Cassini's arrival thanks to earlier images that showed the moon has an unusually bright surface and craters that look like they are blanketed in snow. Nevertheless, the discovery of the ice plumes – initially made when Cassini flew through one – was a spectacular confirmation that Enceladus is an active world.

With a diameter of 504 kilometres (313 miles) and a rock/ice composition, Enceladus should have frozen solid billions of years ago, like many of its neighbours in the Saturnian system. But tidal forces caused by a tug of war between Saturn and a larger moon, Dione, keep the moon's interior warm and active, making it a prime target in the hunt for life in the Solar System.

While much of the water ice falls back to cover the surface, a substantial amount escapes from the weak gravity and enters orbit around Saturn. Here it spreads out to form the E Ring – the outermost and sparsest of Saturn's major rings.

Callisto
The most cratered world

Mass: 1.1×10^{23}kg (2.4×10^{23}lbs)
Diameter: 4,821km (2,996 miles)
Parent planet: Jupiter
Discovered: 1610, Galileo Galilei

The outermost of Jupiter's Galilean moons, Callisto is the third-largest moon in the Solar System and is only slightly smaller than Mercury. Its main claim to fame is the title of most heavily cratered object in the Solar System; its dark surface is covered in craters down to the limit of visibility, the deepest of which have exposed fresh ice from beneath and scattered bright 'ejecta' debris across the surface.

Callisto owes its cratered surface to its location in the Jupiter system – the giant planet's gravity exerts a powerful influence, disrupting the orbits of passing comets and often pulling them to their doom, most spectacularly demonstrated in the 1994 impact of Comet Shoemaker-Levy 9.

Jupiter's larger moons are directly in the firing line and end up soaking up more than their fair share of impacts, but Callisto's inner neighbours – influenced by greater tidal forces – have all experienced geological processes that wiped away most of their ancient craters. Callisto's surface, however, has remained essentially unchanged for more than 4.5 billion years, developing its dense landscape of overlapping craters across aeons.

1 Inner rings
The broad Cassini Division separates the A Ring from the inner B, C and D Rings.

2 A Ring
Saturn's brightest ring is the A Ring, divided by the narrow Encke Gap. The ultra-fine F Ring runs around its outer edge.

3 E Ring
The E Ring is a broad, largely transparent ring of scattered icy particles.

4 Orbit of Enceladus
Enceladus' orbit around Saturn coincides with the densest part of the E Ring.

5 Geyser faults
Tidal forces create an ocean that erupts as icy plumes along weak fault lines in the southern hemisphere.

6 Fragmented material
Ice fragments trailing behind Enceladus are ground down into ever-finer particles as they orbit Saturn.

UNDERSTANDING THE SOLAR SYSTEM

Dactyl The asteroid moon

Mass: 1.1 x 10²⁰kg (2.4 x 10²⁰lbs) **Diameter:** 504km (313 miles)
Parent planet: Saturn **Discovered:** 1789, William Herschel

243 Ida's moon is tiny, just 1.6 kilometres (0.99 miles) on its longest axis. Thanks to the larger asteroid's weak gravity, Dactyl is unlikely to be an object captured into orbit.

Ida is a major member of the Koronis family of over 300 asteroids, all of which share similar orbits. The family is thought to have formed 1 or 2 billion years ago during an asteroid collision. Dactyl could be a smaller fragment of debris from the collision that ended up in orbit around Ida, but there is a problem – computer models suggest Dactyl would almost certainly be destroyed by an impact from another asteroid. So how can it be over a billion years old?

One theory is that the Koronis family is younger than it appears, and Ida's cratering is due to a storm of impacts triggered in the original break-up. Another theory is that Dactyl has suffered a disrupting impact, but has pulled itself back together in its orbit – which might explain its spherical shape.

1 Direction of orbit Dactyl orbits in the same direction as Ida's rotation.

2 Encounter distance Dactyl was about 90 kilometres (56 miles) from Ida during Galileo's flyby.

3 Close approach Computer models show that Dactyl can come no closer to Ida than 65 kilometres (40 miles) for its orbit to remain stable.

4 Range of possibilities These tracks show a range of potential orbits that would fit Galileo's observations.

Iapetus
The walnut

Mass: 1.8 x 10²¹kg (4.0 x 10²¹lbs) **Diameter:** 1,469km (913 miles)
Parent planet: Saturn **Discovered:** 1671, Giovanni Cassini

Iapetus has two distinct claims to a place in any list of weird satellites. The first became obvious when it was discovered in 1671 – it is much dimmer when seen on one side of its orbit compared to the other. Its leading hemisphere – the half that faces 'forwards' as it orbits Saturn – is dark brown, while its trailing hemisphere is light grey. One early theory to explain the colour difference was that the leading side is covered in dust generated by tiny meteorite impacts on small outer moons, which spirals towards Saturn.

However, images from Cassini reveal a more complex story. Most of the dark material seems to come from within Iapetus, left behind as dark 'lag' when dust-laden ice from the moon's surface sublimates – turns from solid to vapour. The process was likely started by dust from the outer moons accumulating on the leading hemisphere, but once it began, the tendency of the dark surface to absorb heat has caused a runaway sublimation effect.

Iapetus is also ringed by a mountainous equatorial ridge that is 13 kilometres (eight miles) high and 20 kilometres (12 miles) wide, giving the moon its distinctive walnut shape. The origins of this ridge are puzzling – some theories suggest it is a 'fossil' from a time when Iapetus span much faster and bulged out at the equator, while others think it could be debris from a ring system that once encircled the moon and collapsed onto its surface.

Nereid
Neptune's boomerang

Mass: 3.1 x 10¹⁹kg (6.8 x 10¹⁹lbs)
Diameter: 340km (211 miles)
Parent planet: Neptune
Discovered: 1949, Gerard Kuiper

Nereid was the second moon found to orbit Neptune, and its claim to fame arises from its extreme orbit. Nereid's distance from Neptune ranges between 1.4 million and 9.7 million kilometres (870,000 and 6 million miles). This orbit is usually typical of captured satellites – asteroids and comets swept up into highly eccentric orbits by the gravity of the giant outer planets – but Nereid's unusually large size suggests a rather more interesting story.

Evidence from Voyager 2's 1989 flyby suggests that Triton was captured into orbit from the nearby Kuiper Belt. Triton would have disrupted the orbits of Neptune's original moons, ejecting many of them. But many astronomers believe Nereid could be a survivor, clinging on at the edge of Neptune's gravitational reach.

STRANGEST MOONS
IN THE SOLAR SYSTEM

Io The cold inferno

Mass: 8.93 x 10²²kg (1.97 x 10²³lbs) **Diameter:** 3,643km (2,264 miles)
Parent planet: Jupiter **Discovered:** 1610, Galileo Galilei

Io is the innermost of the four giant Galilean moons that orbit the Solar System's largest planet, Jupiter. But while the outer three are – at least outwardly – placid, frozen worlds of rock and ice, Io's landscape is a virulent mix of yellows, reds and browns, full of bizarre and ever-changing mineral formations created by sulphur that spills onto its surface in many forms. Io is the most volcanic world in the Solar System. Io's strange surface was first observed during the Pioneer space probe flybys of the early 1970s, but its volcanic nature was only predicted weeks before the arrival of the Voyager 1 mission in 1979.

The moon is caught in a gravitational tug of war between its outer neighbours and Jupiter itself, and this prevents its orbit from settling into a perfect circle. Small changes in Io's distance from Jupiter – less than 0.5 per cent variation in its orbit – create huge tidal forces that pummel the moon's interior in all directions. Rocks grinding past one another heat up due to friction, keeping the moon's core molten and creating huge subsurface reservoirs of magma.

While the majority of Io's rocks are silicates similar to those on Earth, these have relatively high melting points, and so are mostly molten in a hot magma ocean that lies tens of kilometres below the surface – most of Io's surface activity, in contrast, involves sulphur-rich rocks that can remain molten at lower temperatures.

Together these two forms of volcanism have long since driven away any icy material that Io originally had, leaving a world that is arid and iceless despite an average surface temperature of -160 degrees Celsius (-256 degrees Fahrenheit).

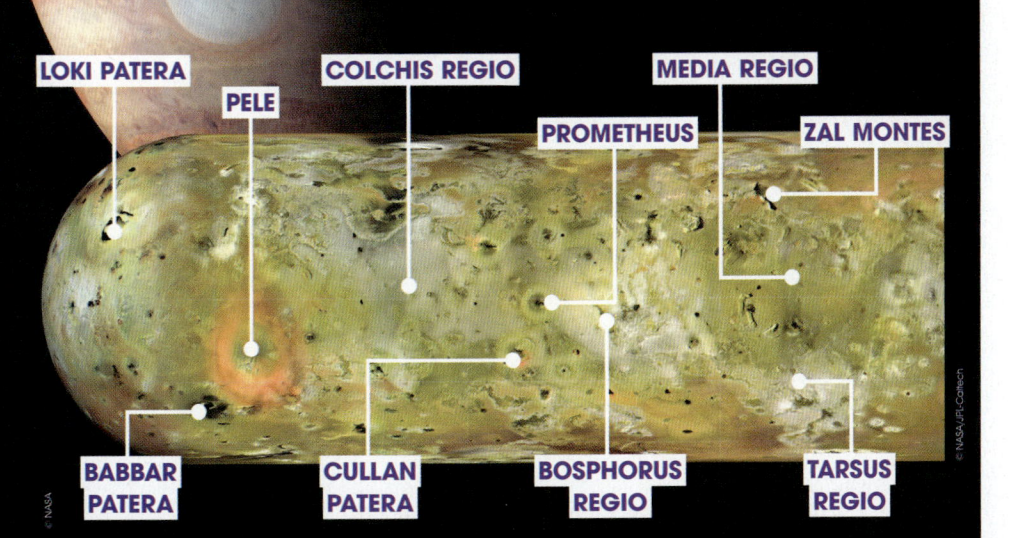

LOKI PATERA
PELE
COLCHIS REGIO
MEDIA REGIO
PROMETHEUS
ZAL MONTES
BABBAR PATERA
CULLAN PATERA
BOSPHORUS REGIO
TARSUS REGIO

Hyperion
The spongy satellite

Mass:
5.6 x 10¹⁸kg (1.2 x 10¹⁹lbs)
Diameter:
270km (168 miles)
Parent planet: Saturn
Discovered: 1848, William Bond, George Bond and William Lassell

Hyperion is the strangest-looking satellite in the Solar System, its surface resembling a sponge or coral with deep, dark pits rimmed by razor-sharp ridges of brighter rock and ice. But that's not the only thing that's strange about Hyperion: it was the first non-spherical moon to be discovered and has a distinctly eccentric orbit.

Rather than matching its rotation to its orbital period, it spins in a chaotic pattern, with its axis of rotation wobbling unpredictably. Like all moons in the outer Solar System, it's mostly made of water ice, but its surface is unusually dark. When Cassini flew past it measured its density to be 55 per cent that of water – its interior is mostly empty space.

One popular theory to explain these weird features is that Hyperion is the surviving remnant of a larger satellite that once orbited between Titan and Iapetus, and which was largely destroyed by a collision with a large comet. Material that survived in a stable orbit then came together again to create Hyperion as we know it.

UNDERSTANDING THE SOLAR SYSTEM

1 Weak sunlight
Titan's distance from the Sun and its thick atmosphere mean that the surface receives about one per cent of the sunlight that Earth receives.

2 Tiny Sun
From Saturn, the Sun is ten per cent of the size as seen from Earth.

3 Methane loss
Methane evaporates from lakes back into the atmosphere.

4 Giant planet
Saturn's huge bulk is largely hidden by a hazy atmosphere.

5 Methane lakes
Methane rains out of the atmosphere at the winter pole to form large lakes.

6 Landscape runoff
Methane rainfall onto highland areas runs downhill and collects in methane lakes.

7 Solid methane
Methane frosts coat a landscape of rock and water ice.

Titan The second Earth

Mass: 1.3×10^{23}kg (2.9×10^{23}lbs) **Diameter:** 5,150km (3,200 miles)
Parent planet: Saturn **Discovered:** 1655, Christiaan Huygens

Saturn's largest moon, Titan is unique in the Solar System as the only satellite with a substantial atmosphere of its own – a discovery that frustrated NASA scientists when images from the Voyager probes revealed only a hazy orange ball. The Cassini orbiter was fitted with infrared and radar instruments that pierced the opaque atmosphere, revealing a softened landscape of rivers and lakes that is unlike any other world in the Solar System except for Earth. Despite being larger than Mercury, Titan can only hold onto its thick atmosphere because of the deep cold. Found some 1.4 billion kilometres (0.9 billion miles) from the Sun, the moon's average surface temperature is a freezing -179 degrees Celsius (-290 degrees Fahrenheit).

Titan's atmosphere is dominated by the inert gas nitrogen – also the major component of Earth's air – but it gets its distinctive colour, opaque haze and clouds from a relatively small proportion of methane. Amazingly, conditions on Titan are just right for methane to shift between its gaseous, liquid and solid forms, generating a 'methane cycle' rather similar to the water cycle that shapes Earth's climate. In cold conditions methane freezes onto the surface as frost and ice. In moderate temperatures it condenses into liquid droplets and falls as rain that erodes and softens the landscape before accumulating in lakes, while in warmer regions it evaporates and returns to the atmosphere.

Titan experiences changing seasons very similar to those on our planet, though its year is 29.5 Earth years. Temperatures at the winter pole seem to favour rainfall, so the lakes migrate from one pole to the other over each Titanian year. With all this activity, Titan is an intriguing target in the search for extraterrestrial life, though most biologists find it hard to envision organisms that could exist in such harsh and chemically limited conditions, and most agree that Titan's watery inner neighbour Enceladus offers more promising prospects for life.

Measuring up Titan
■ Titan ■ Earth

SURFACE GRAVITY
- 0.14
- 1.0

Day length in Earth days
- 16
- 1

Orbital period in Earth days
- 16
- 1

Temperature   0°C
- -179°C (-290°F)
- 14°C (57°F)

Atmospheric pressure
- 1.5 bar
- 1.0 bar

Atmospheric composition
- 95% nitrogen, 5% methane
- 78% nitrogen, 21% oxygen, 1% argon

STRANGEST MOONS IN THE SOLAR SYSTEM

Miranda The chiselled satellite

Mass: 6.6 x 10^{19}kg (1.5 x 10^{20}lbs) **Diameter:** 470km (292 miles)
Parent planet: Uranus **Discovered:** 1948, Gerard Kuiper

Miranda is one of the strangest worlds in the Solar System. Voyager images revealed an extraordinary patchwork of terrains. Some parts are heavily cratered and some relatively uncratered – indicating their youth, as they have been less exposed to bombardment.

An early theory to explain Miranda's appearance is that it is a Frankenstein world – a collection of fragments from a predecessor moon that coalesced in orbit around Uranus. Astronomers wondered whether Miranda's predecessor might have been shattered by an interplanetary impact, and whether this event might be linked to Uranus' own extreme tilt. Further studies, however, have shown that such a theory comes up short when trying to explain Miranda's mix of surface features, and the right kind of impact is unlikely. Instead it seems plausible that tidal forces are to blame.

Today Miranda follows a near-circular orbit, but its past orbit was in a 'resonant' relationship with larger moon Umbriel. This brought the two moons into frequent alignments that pulled Miranda's orbit into an elongated ellipse that experienced extreme tidal forces. Pushed, pulled and heated from within, its surface fragmented and rearranged itself before the moons moved and Miranda's activity subsided.

Mimas The real-life Death Star

Mass: 3.8 x 10^{19}kg (8.4 x 10^{19}lbs) **Diameter:** 396km (246 miles)
Parent planet: Saturn **Discovered:** 1789, William Herschel

When NASA's Voyager space probes sent back the first detailed images of Mimas in the 1980s, scientists and the public were shocked by its resemblance to the Death Star space station from the Star Wars films. A huge crater – named after William Herschel, who discovered the moon in 1789 – dominates one hemisphere, and is almost the exact size and shape of the planet-killing laser dish dreamt up by George Lucas many years before. But Mimas has more to offer than pop-cultural references.

Mimas is the innermost of Saturn's substantial moons – orbiting closer than Enceladus but further out than Pan and Atlas – and with a diameter of just 396 kilometres (246 miles), it's the smallest object in the Solar System known to have pulled itself into a spherical shape from its own gravity. Some larger Solar System objects haven't quite managed this, and most astronomers agree that it's only possible for Mimas because of the moon's low density – just 15 per cent greater than water.

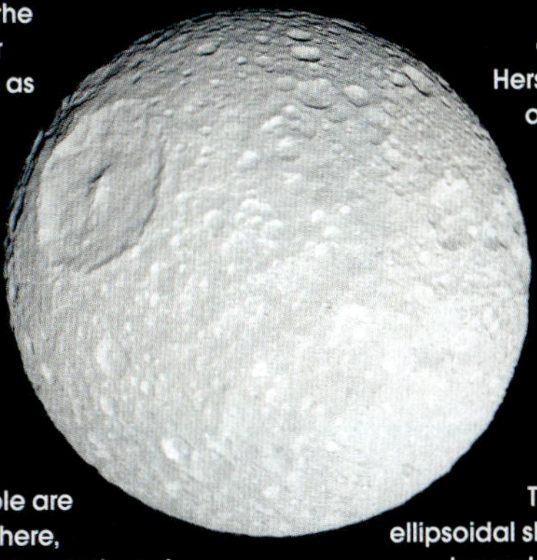

Death Star crater
If Mimas was scaled up to the size of Earth, its giant crater Herschel would be as wide as Australia. The crater was originally far deeper.

Towering peak
Herschel's central peak rises six kilometres (3.7 miles) above the crater floor – it's as high as Mount Kilimanjaro.

Crater variation
Craters around its south pole are half the size of those elsewhere, suggesting this was probably resurfaced with fresh ice early in Mimas' history.

Cracked surface
Deep chasms were likely created in the formation of Herschel. Scaled up to the size of Mars, they would rival the Valles Marineris canyons.

Colour changes
Enhanced-colour images show slight variations in the surface: a slightly greenish hue overlaid with blue around Herschel.

Rugby ball core
Tidal forces have given it an ellipsoidal shape, with the axis pointing towards Saturn ten per cent longer than the axis from pole to pole.

Pan and Atlas
The flying saucers

Mass: 4.9 x 10^{15}kg (1.1 x 10^{16}lbs) and 6.6 x 10^{15}kg (1.6 x 10^{16}lbs)
Diameter: Average of 28km (17 miles) and 30km (19 miles)
Parent planet: Saturn
Discovered: 1990 and 1980, Voyager 2

Saturn's moons Atlas and Pan are the smallest moons in the Solar System. However, despite their size their influence can be seen clearly from Earth in the form of the prominent 'gap' they create in the planet's ring system.

These two tiny worlds are perhaps the best known examples of shepherd moons – small satellites that orbit in or around the ring systems of the giant planets. As the name suggests, when coupled with the influence of distant outer moons, such satellites help to herd the particles orbiting in the ring system together while 'clearing out' others. Pan is responsible for creating the Encke Gap, a prominent division in Saturn's bright A Ring, while Atlas orbits just outside the A Ring.

The most intriguing property of both worlds is their smooth shape, resembling a walnut or a flying saucer. Experts believe the moons are blanketed in small particles swept up as they keep the space between the rings clear. As most of the particles orbit in a plane one kilometre (0.6 miles) thick, they tend to pile up around each moon's equator, building a distinctive equatorial ridge.

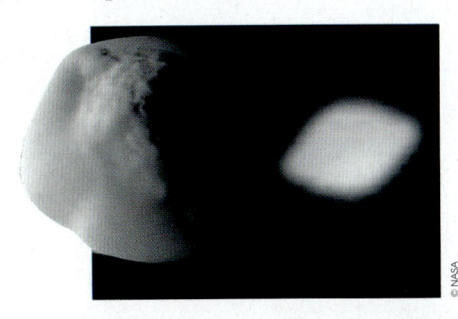

UNDERSTANDING THE SOLAR SYSTEM

Does Earth have a second moon?

There's an asteroid tracking our planet's orbit around the Sun, and astronomers have been surprised by its composition

EARTH'S SECOND MOON

We all know that Earth only has one moon, and if you want proof of that, you just need to peer at the night sky. Can you see another? Astronomers are sticking to their guns by saying the Moon is Earth's only natural satellite. Case closed… or so you'd think. In 2016, however, astronomers using the asteroid-hunting Pan-STARRS telescope in Hawaii discovered a rock orbiting the Sun while also repeatedly looping around Earth. Celestial objects such as these are called quasi-satellites, and while four others have been identified since 2004, this one was the closest and most stable ever seen.

What makes it particularly interesting is its potential origin, although a recent breakthrough has gone a long way towards clearing up the five-year mystery over what this object could be. The smart money is now on it actually being a fragment of our Moon. In that sense you could say it's Earth's second moon – as some have indeed dubbed it – though it's not quite on that level, if truth be told. Still, it's no less intriguing, and astronomers are keen to discover more.

One of the people leading studies into the object, which is being called Kamoʻoalewa – a Hawaiian name that roughly means 'oscillating celestial fragment' – is University of Arizona planetary sciences graduate student Ben Sharkey. For the past five years he has dedicated much of his time and energy into finding out the origin of the celestial body, with interest piquing following the recent publication of his team's academic paper in the scientific journal Communications Earth & Environment.

Getting this far has certainly been no easy task. The eccentric entity is no more than 58 metres (190 feet) in diameter, which makes it roughly the size of a Ferris wheel. What's more, the orbiting object is 4 million times fainter than any human could see with the naked eye and about 40 times farther out than the Moon. At its greatest distance it is 25 million miles away, and it gets no closer to Earth than 9 million miles. But there it is, in orbit around Earth and the Sun, albeit following a rather strange path.

"Kamoʻoalewa is kind of weaving inside and outside of Earth's orbit as both it and our planet go around the Sun," Sharkey explains, this behaviour resulting from the Sun and Earth's gravitational pulls competing with one another. "If you look at that from the perspective of Earth, it looks like a cork scurrying around the

UNDERSTANDING THE SOLAR SYSTEM

Kamoʻoalewa by numbers

25 million
The farthest Kamoʻoalewa gets from Earth, in miles

9 million
The closest Kamoʻoalewa gets to Earth, in miles

2016
Year it was discovered

365.9
Orbital period, in days

45 to 58 metres
The estimated size of Kamoʻoalewa

30
Kamoʻoalewa's rotational period, in minutes

Nine
Number of asteroids found to have lunar origins

4 million
times fainter than the faintest star visible with the naked eye

500 to 100,000
Years ago it may have broken off the Moon

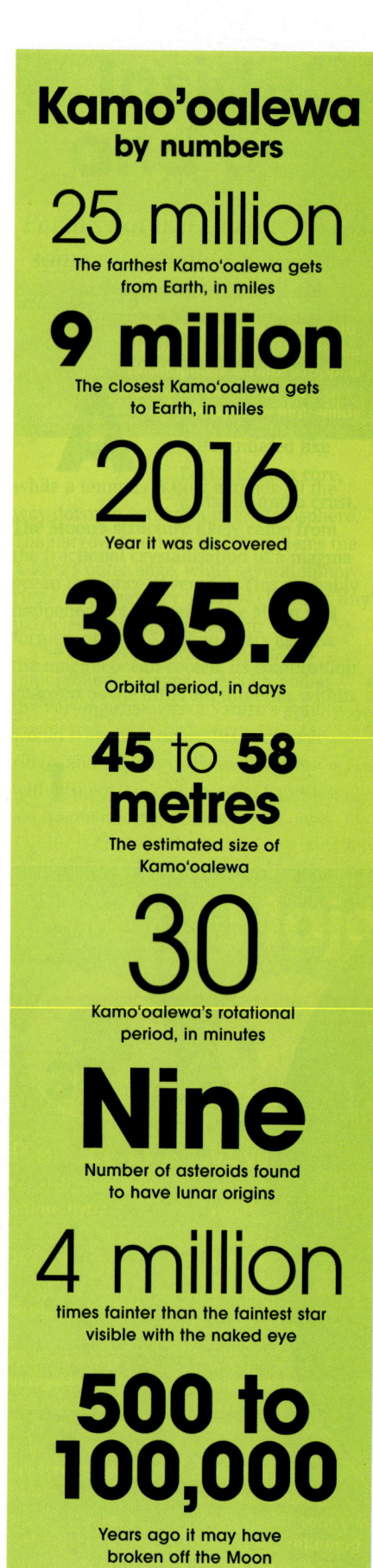

planet in an orbit that doesn't close at any point. It never repeats the same loop in exactly the same way."

The celestial object's odd corkscrew-like orbit certainly caught astronomers' imagination. "We could tell this object had a unique orbit, so there was an immediate interest in characterising it to determine what it's made of and how it is spinning," Sharkey says. "As a student, it seemed like a challenging and interesting project to sink some time into, so we collected a bunch of different kinds of data over a long period of time." Such diligence has been reaping rewards.

There's no doubt that the Arizona study has progressed slowly, but there's been no way of speeding it up. It's only been possible to observe Kamoʻoalewa for a few weeks every April, when it was illuminated by the Sun, but astronomers have nevertheless been able to observe the object sufficiently to draw some conclusions. They have been primarily making use of the Large Binocular Telescope on Mount Graham in southern Arizona. The observatory's huge nine-metre (28-foot) mirrors allowed a much closer look at Kamoʻoalewa, with observations beginning in 2017.

Backed by data from the Lowell Discovery Telescope, which is a five-hour drive away in Flagstaff and funded by NASA's Near-Earth Object Observations Program, the astronomers have been able to monitor Kamoʻoalewa's faint infrared signature. The team would make observations, then plan ahead for the following year's window of opportunity, building up a solid bank of knowledge. "Asteroids like this are essentially darker than the sky and fainter than the background glow, so you need large telescopes and infrared instruments to make these detections," Sharkey explains. Slowly but surely, the team's observations and analyses pointed them towards Kamoʻoalewa being a lost piece of the Moon.

There's a big reason why this is a major deal. If Kamoʻoalewa is indeed a lunar fragment then it would be the only known asteroid with a lunar origin. It also opens up a whole new line of inquiry that explores how the piece came to break away and when it happened. Of course,

EARTH'S
SECOND MOON

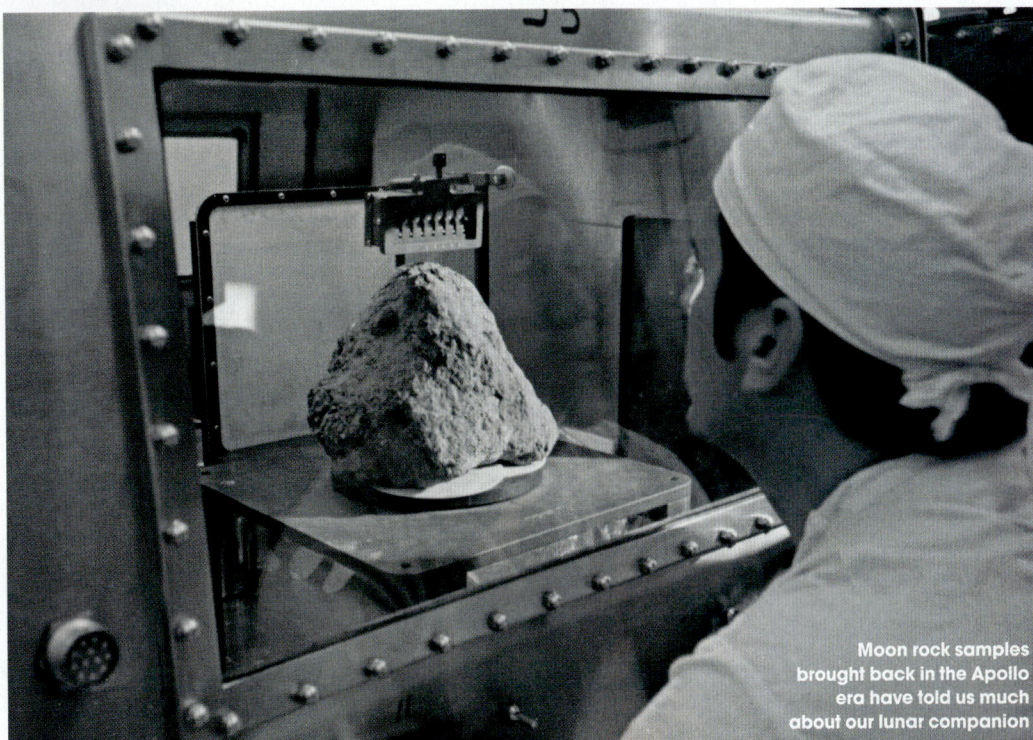

Rock samples collected from the Moon during the Apollo 14 mission compared almost perfectly to the data gathered about Kamoʻoalewa

Moon rock samples brought back in the Apollo era have told us much about our lunar companion

'Big Bertha', the third-largest lunar sample ever collected

> "Kamoʻoalewa is kind of weaving inside and outside of Earth's orbit as both it and our planet go around the Sun"
> **Ben Sharkey**

such events have occurred many times in the past. "You can see craters on the Moon with your own eyes," explains Sharkey. But we think of most asteroids as coming from the asteroid belt.

"We know material from the Moon has been ejected in the past, some of it coming to Earth as meteorites," Sharkey says. "Hundreds of lunar meteorites have been collected on Earth, so impacts on the Moon are not uncommon. But the question is whether we can find a link in the processes. Is there a possibility that material from the Moon that hasn't hit Earth is out there, yet to be found? We could tell a lot about how material has mixed or moved around the Solar System with time."

Sharkey and his team knew early on that Kamoʻoalewa wasn't a standard near-Earth

UNDERSTANDING THE SOLAR SYSTEM

Plotting Kamo'oalewa's orbit
How the asteroid became Earth's moon-like buddy

1 Orbiting the Sun
Kamo'oalewa is also known as 2016 HO3, and just like Earth it orbits the Sun, taking just over a year to do so.

2 Staying close
Kamo'oalewa also circles Earth, and it has become a stable quasi-satellite. It doesn't venture too far away as both the asteroid and Earth orbit the Sun.

3 Weaving in and out
The asteroid spends roughly half of the time closer to the Sun than Earth; the objects leapfrog so that the other half of the time it is farther away.

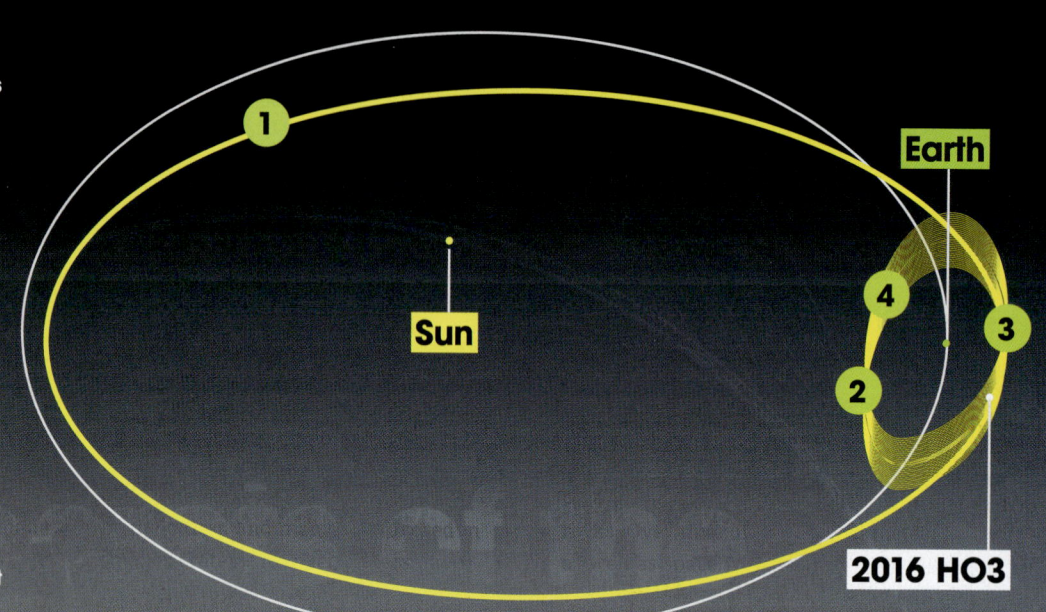

4 Gravitational pull
This unusual orbit is due to Earth's gravity being strong enough to hold onto the asteroid. It doesn't venture farther than 100 times the distance of Earth to the Moon.

Pan-STARRS is the world's leading near-Earth object discovery telescope; it picked up on Kamo'oalewa in 2016

asteroid. During observations it wasn't reflecting brightly in particular infrared frequencies, even though it was made of common silicates in the same way as other asteroids. Sharkey sought to match data gathered on Kamoʻoalewa with the light reflecting off other near-Earth asteroids. But Kamoʻoalewa's dimmer reflection pointed to it being composed of a different material. Researchers just needed to be totally sure there had not been a mistake.

"The spectral signature of some other asteroids can actually look very similar to what Kamoʻoalewa looks like at visible wavelengths," Sharkey says. "We had this expectation that it would basically reflect infrared light in a similar way to how it did in the visible, but it displayed a redder reflectance spectrum. We tried to run through different scenarios and think about the different ways to change how asteroids reflect light. This doesn't have to do with just the material it's made of. It could be to do with different textures of the surface – scuffing can make something appear darker or change its colour, for instance, altering how asteroids are reflecting light.

"But we tried to say if we take a more typical asteroid and apply this effect, can we come up with a satisfying answer and achieve a reflection identical to what we saw with Kamoʻoalewa? And our argument was that we can't get anything else to work. Then we saw that there's a source of material that has a very good match to this spectrum which is nearby: space-weathered lunar silicates." It opened up a winning line of inquiry.

As luck would have it, one of Sharkey's PhD advisors had studied samples collected from the Fra Mauro formation in the lunar highlands which had been brought back to Earth by the Apollo 14 mission in 1971. "It was a kind of funny moment because he just sort of mentioned it randomly, and it was like, 'hey, you should probably think about this comparison too'," Sharkey laughs. By comparing Kamoʻoalewa's pattern of reflected light to those lunar

Denise Hung and Dave Tholan of the University of Hawaii took this image of Kamoʻoalewa on 10 June 2016

samples, a near-perfect match was found. This suggested that Kamoʻoalewa indeed originated from the Moon, with the spectral characteristics being consistent with silicate material showing a high degree of space weathering, such as solar wind particles or micrometeorite bombardment.

Even so, Sharkey and his advisor, University of Arizona associate professor of lunar and planetary sciences Vishnu Reddy, harboured doubts. Although a second set of data was obtained in 2019, they were frustrated that they couldn't observe Kamoʻoalewa in 2020 because the COVID-19 pandemic caused the Large Binocular Telescope to be shut down. Fortunately, it was back up and running in April this year, allowing another observation, and this finally convinced the team that they were correct.

But Kamoʻoalewa isn't going to hang around. According to the study's coauthor, University of Arizona planetary sciences professor Renu Malhorta, who has been studying the orbit, it will wave goodbye in about 300 years time, once it frees itself from the gravitational ropes that are keeping it around Earth. "It will not remain in this particular orbit for very long, only about 300 years in the future," Malhorta affirms, adding that "it arrived in this orbit about 500 years ago". But what exact path it will take when it goes on its lonely adventure is not entirely certain. What we do know, however, is that it's

The theories

Astronomers have considered a few explanations for Kamoʻoalewa's origin

A captured asteroid

The study's authors examined the possibility that Kamoʻoalewa was captured in its Earth-like orbit from the general population of near-Earth objects. But the paper says that simulations of such a scenario don't match the low eccentricity and inclination displayed by Kamoʻoalewa.

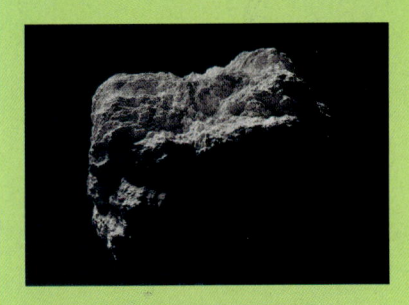

Fragment of an asteroid

There are two known asteroids that orbit the Sun at the Earth-Sun Lagrangian points L4 and L5. They have similar orbits to Earth's, and it could be that Kamoʻoalewa is a split-off piece of one of them, drawn from a quasi-stable population that has yet to be found.

A piece of the Moon

The recent study of Kamoʻoalewa strongly suggests that the asteroid is an ejected fragment of the Moon, probably caused by an impact. This is supported by the reflectance spectrum of Kamoʻoalewa being a near-perfect match in comparison to rock samples returned from the Moon.

> "We could tell this object had a unique orbit, so there was an immediate interest in characterising it"
> **Ben Sharkey**

UNDERSTANDING THE SOLAR SYSTEM

An image of the lunar surface taken on the Apollo 14 mission

> "It will not remain in this particular orbit for very long, only about 300 years in the future; it arrived in this orbit about 500 years ago"
>
> **Renu Malhorta**

gravitationally bound to the Sun rather than Earth, and that's why it can't actually be classed as a moon, even though it's made up of a bit of one.

"Orbital dynamics are outside my normal expertise," says Sharkey, "but one of the big challenges is tracing the exact path of an object like this beyond a few hundred years. Your uncertainties just become so large that it's really difficult to say anything with confidence besides the sort of general ideas about the state of the object. Beyond 300 years or so, it's a question of the exact path it takes, but it's not going to shoot off in some new direction. It's going to be a more gentle progression than that."

This still gives plenty of time to make more observations, and while there's no chance anyone is going to be thinking of landing a human on this particular 'moon', a sample-return mission to Kamoʻoalewa is being planned by the China National Space Administration, with a launch pencilled in for 2025. It's also entirely possible that Kamoʻoalewa is not alone, and the orbits of three other near-Earth objects could even be linked. "I feel at this point that anything could be a surprise, or not a surprise, and the honest answer is that we really don't know," Sharkey says. "I think that's part of why this kind of study adds excitement, at least from my perspective. But we didn't really have a good handle on the question of what Kamoʻoalewa was beforehand and whether other objects could be related in similar ways. The next step is to keep asking that question."

Kamoʻoalewa is roughly the size of Italy's Leaning Tower of Pisa

David Crookes
Science and technology journalist

David has been reporting on space, science and technology for many years, has contributed to many books and is a producer for BBC Radio 5 Live.

DISCOVER THE PAST, PRESENT AND FUTURE OF SPACE EXPLORATION

From the formative years of Sputnik through to modern-day innovations like Perseverance, embark on a journey across the history of humanity's missions into space, and glimpse what is in store for the future.

ON SALE NOW

Ordering is easy. Go online at:
magazinesdirect.com
Or get it from selected supermarkets & newsagents

UNDERSTANDING THE SOLAR SYSTEM

Escape to Titan

When the Sun scorches Earth, a tiny moon in orbit around the ringed giant Saturn is our next home

TITAN

Our planet may have survived for 4.5 billion years, but humanity faces some major threats. There's always the possibility that an asteroid will wipe us out, just as one did for the dinosaurs some 65 million years ago. We could be engulfed by a gamma-ray burst or disrupted by a wandering star. There are also dangers closer to home, from volcanoes to nuclear war. Even supposing humans manage to survive all these threats, we can still say for certain that life here on Earth will eventually be no more. In around 5 billion years from now the Sun will undergo a massive change that will fundamentally alter our Solar System. It will cause the end of not only all life here on Earth, but possibly the entire planet, and we will have no choice other than to find somewhere else to live.

Astronomers have been looking at the possibilities of colonising other planets for years. Mars currently tops the list of destinations, with NASA working hard to develop the capabilities needed to send humans to the Red Planet in the 2030s. Yet some scientists are taking a much longer view. Rather than looking towards the terrestrial planets for our new home, they say humans will one day have to relocate to the outer Solar System if they want to survive. As it currently stands, sending scores of humans to live beyond the asteroid belt is out of the question. The four gas giants are utterly unsuitable for life, and the moons of the outer Solar System are well outside of the habitable zone – the region around the Sun where the atmospheric pressure is able to support liquid water, making conditions for life as we know it 'just right'.

Yet things can and will change. The Sun is getting gradually warmer, and it will eventually become so hot that it will boil off Earth's oceans. This will happen sooner than we think. "In around a billion years' time, Earth will probably no longer be habitable for humans," says Benjamin Charnay, a research associate at the French National Centre for Scientific Research from the Observatoire de Paris. "The increasing solar insulation means Earth will either evaporate all of its oceans or lose them by the atmospheric escape of hydrogen."

But that will only be the start. Even if humans do somehow survive the mass loss of water on Earth, the next thing to be affected would be the current habitable zone. At some stage the Sun's hydrogen supplies at its core are going to deplete, and gravity will take over. Nuclear fusion – the energy-generating process of converting hydrogen into helium – will cease, bringing an end to 10 billion years of stability. As the Sun's core collapses, helium will fuse into carbon and the Sun will bloat into a red giant, 256 times its original size. "The Sun will swell out to beyond the orbit of Earth," says Dr Christopher McKay, a planetary scientist at NASA Ames Research Center. The effects of this will be devastating for both Earth and the inner Solar System, and at this stage staying on our planet will not be an option.

After all, the outer layers of the Sun will now be at escape velocity and peeling away. Mercury and Venus will be engulfed, and the orbits of the planets will be widening due to a weakened gravitational pull.

The Cassini-Huygens mission sent back up-close views of the Saturnian moon

> "In around a billion years' time, Earth will probably no longer be habitable for humans"
> **Benjamin Charnay**

Titan's water ice holds key ingredients necessary for life – they just need heating

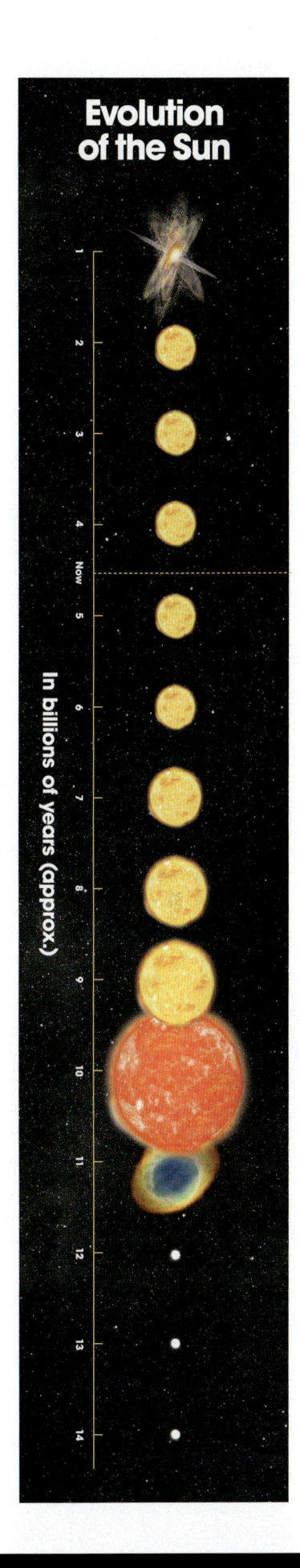

Evolution of the Sun

In billions of years (approx.)

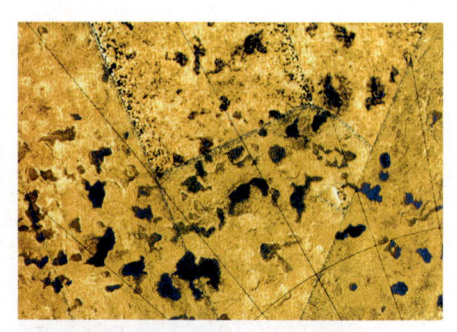

UNDERSTANDING THE SOLAR SYSTEM

This radar image taken by the Cassini spacecraft shows empty and liquid-filled depressions on Titan

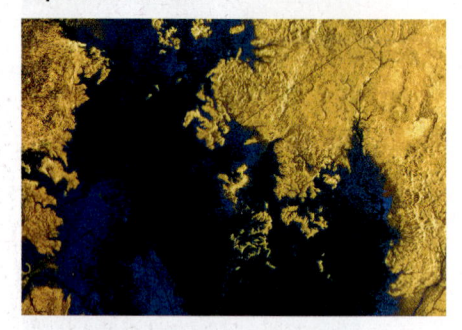

A close-up radar image of Ligeia Mare, the second-largest known body of liquid hydrocarbons on Titan

A potential home
When the habitable zone shifts, Titan could be our first destination

Titan today

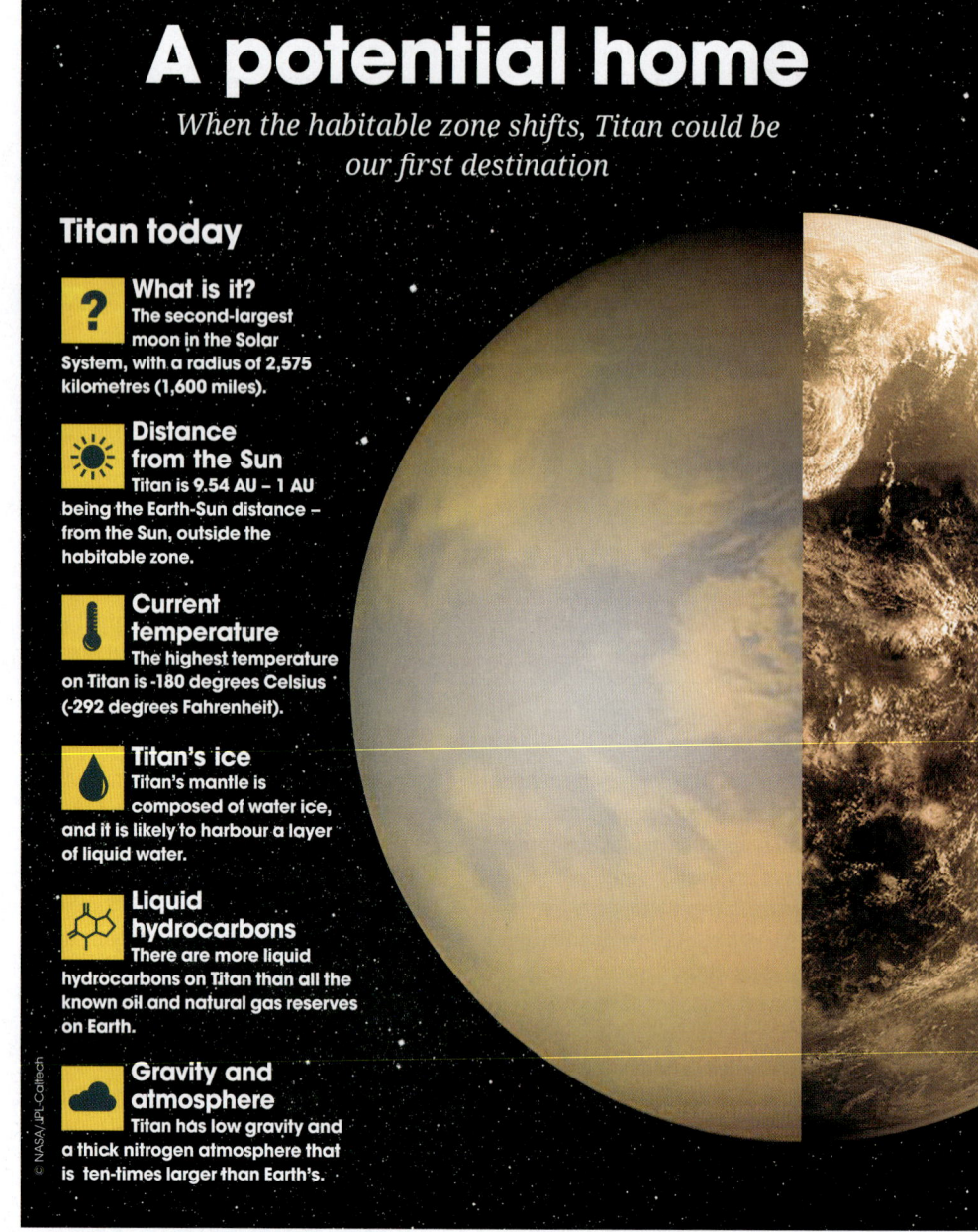

What is it?
The second-largest moon in the Solar System, with a radius of 2,575 kilometres (1,600 miles).

Distance from the Sun
Titan is 9.54 AU – 1 AU being the Earth-Sun distance – from the Sun, outside the habitable zone.

Current temperature
The highest temperature on Titan is -180 degrees Celsius (-292 degrees Fahrenheit).

Titan's ice
Titan's mantle is composed of water ice, and it is likely to harbour a layer of liquid water.

Liquid hydrocarbons
There are more liquid hydrocarbons on Titan than all the known oil and natural gas reserves on Earth.

Gravity and atmosphere
Titan has low gravity and a thick nitrogen atmosphere that is ten-times larger than Earth's.

"Even if it survives, Earth will be inside the Sun's atmosphere," McKay adds. If all of this sounds quite gloomy for the future of humankind, then be assured that it is. "The issue for humans would be to survive to this time," says Charnay, strongly hinting that there is every chance that nobody will be around to see any of it.

Let's suppose that humans do manage to get that far. Where will they be able to go when the Sun has turned a vivid red? The smart money is on a lovely home overlooking Saturn and its stunning system of rings. In the new order of the Solar System, Titan, Saturn's largest moon, is likely to become the number one destination for humans. It won't be easy – the journey to Titan from Earth takes some seven years, which will excessively burden the body and mind of any astronaut – but it could be the perfect escape route that will keep humankind going for many more millions of years.

It may be hard to imagine that a moon which is ten-times further away from the Sun than Earth could possibly become a new human base, but Titan is actually a close match for our planet. In many ways it mimics Earth's primitive state. As such it has proved fascinating for astronomers who have been building up data about the moon since Huygens, an atmospheric entry probe, landed there in 2005 following a seven-year journey as part of the Cassini-Huygens mission. It was the first-ever landing accomplished in the outer Solar System, and it will not be the last by any means. When the day comes that space agencies are seeking to send manned flights to Titan, you can be assured that the technology needed to safely transport people 3.2 billion kilometres (2 billion miles) across the Solar System will be very much in place.

Titan is one of only three worlds in the Solar System with rocky surfaces and thick atmospheres – Venus and Earth being the others. "The thick atmosphere cuts down on radiation, so it is a very neutral environment," says Dr Mike Malaska, a scientist at NASA. It has a gravity that is similar to that of our own natural orbital satellite, making Titan the easiest place to fly and land in the Solar System – something that should help with future colonisation. Astronauts will be able to navigate Titan wearing just warm coats as it benefits from having zero to low pressure, unlike our Moon or Mars. "The Moon and Mars both share the problem that if humans didn't wear spacesuits they

98

TITAN

Future Titan

What is it?
Titan will remain the second-largest moon as Jupiter's moon Ganymede will still exist.

Distance from the Sun
During the red giant phase of the Sun, Titan will be in the habitable zone.

Rising temperatures
Titan's surface temperature could rise to -70 degrees Celsius (-94 degrees Fahrenheit).

Flowing water
With a hotter climate and melted ice, Titan would host large oceans of life-giving water.

Organics will emerge
Titan should have around 100 million years for organics to heat and flourish.

Lovely atmosphere?
The upper atmospheric haze will deplete and methane-based greenhouse effects will kick in.

Bright sunlight can be seen reflecting off Titan's hydrocarbon seas, which are mostly liquid methane and ethane

---

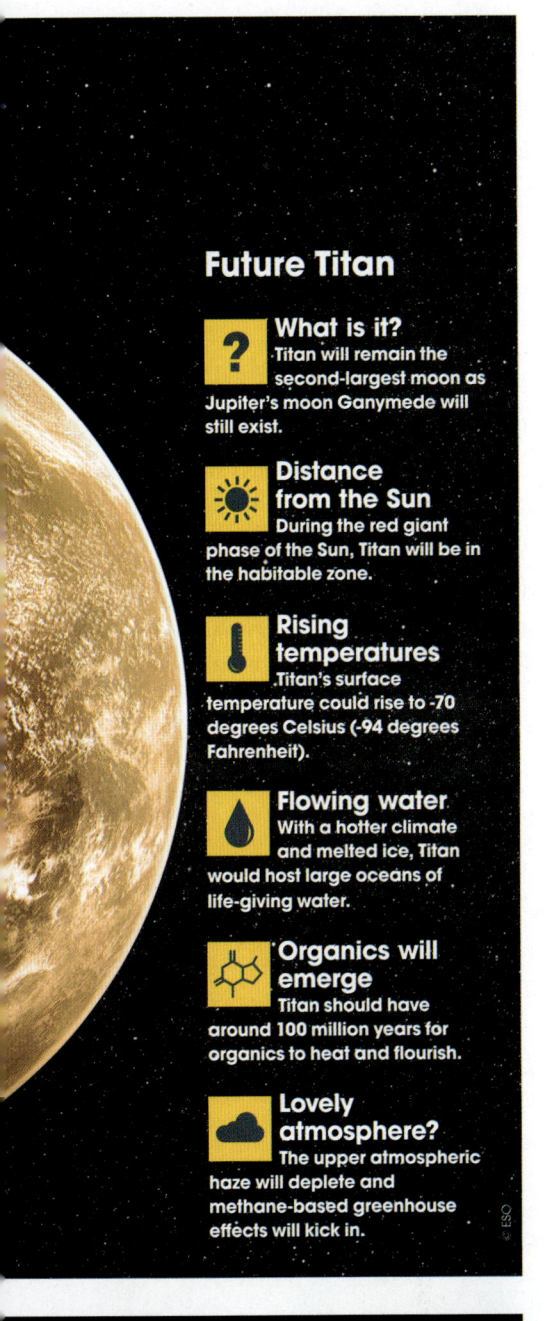

would die rapidly from depressurisation, which the movies like to show as being explosive," says McKay. "On Titan a spacesuit is not required."

Titan has weather, and it is the only body in the Solar System other than Earth to possess surface lakes and seas. It also has river channels, dunes and complex hydrocarbons, along with pebbles of ice that point to an existence of water in the past. Crucially, it has copious organic raw materials. "These would be great for colonialists to use for manufacturing things," adds Malaska. "The diversity of features on the surface suggests that there might be different patches of different types of organics – kind of like the different rock outcrops here on Earth. There might also be outcrops of water ice, so water might be available after heating it."

Indeed, astronomers say that all Titan effectively needs is warming up to make it a viable home. "Temperature is a current problem on Titan," says McKay, as the moon receives one-hundredth of the solar heat we get here on Earth. "At -180 degrees Celsius (-292 degrees Fahrenheit), if you visited today it would feel like plunging into freezing cold water, so humans would have to find a way of keeping very warm. We'd have to wear special spacesuits like the ones divers wear. Yet all this changes once the Sun becomes a red giant."

When the Sun has transformed, Titan will be in the middle of a new habitable zone, which will have moved deeper into the Solar System, taking it as far as the Kuiper Belt. The frozen moons of the outer planets will become far warmer, melting ice into liquid water and allowing life a chance to flourish. As McKay, Ralph Lorenz and Jonathan Lunine wrote in an important paper published back in 1997, Titan would respond well to being given a new lease of life, and humans could benefit greatly from it.

Quite apart from the Sun raising the moon's temperature to -70 degrees Celsius (-94 degrees Fahrenheit), they noted that the surrounding thick, orange haze of Titan's atmosphere would also be depleted. Since the haze currently allows the surface to be unaffected by the increase of solar radiation caused by the Sun being closer and hotter, this would enable a greenhouse effect, creating an environment that would

> **"The Sun will swell out to beyond the orbit of Earth. Even if it survives, Earth will be inside the Sun's atmosphere"**
> Christopher McKay

be suitable for life. Things would begin to slot into place.

What's more, the moon's warmer temperature would also change the composition of the atmosphere. Currently it is made up of 95 per cent nitrogen and five per cent methane and "the air is thicker than Earth by a factor of seven," says McKay. But 5 billion years from now "the luminosity will be large enough to affect the icy crust and liberate a water-ammonium ocean," says Charnay. The effects will be jaw-dropping.

Dr Carrie Anderson, planetary astronomer at the Astrochemistry Laboratory at NASA Goddard Space Flight Center, says the water will melt, mix with the organics and make up amino acids, which contain carbon, nitrogen, oxygen and hydrogen – the basic elements necessary for life. "Titan has all of these," she told an audience at the Library of Congress. "It's just waiting. It's ready to go."

Even so, there are still some doubts: "It would be difficult to produce an oxygen-rich atmosphere because Titan's atmosphere and interior are very reducing," says Charnay. "For instance, there is a lot of methane, which would react to destroy oxygen." There would also be something of a race against time in order for life to form and flourish. Titan will have a window of 'just' 100 million years for life to emerge, for reasons we'll come to in a moment. Anderson believes that this is sufficient time for life to form on Titan, however, making it a viable future home for humans. "At this moment in time the ice will melt in the mantle, and a lot of it should melt, so you should have liquid water. Then we have all those organics just sitting around on the surface just waiting for the Sun to heat them up," she says.

UNDERSTANDING THE SOLAR SYSTEM

It is a prospect that also excites Malaska. "When the Sun evolves into a red giant, Titan will heat up," he says. "The water ice in the crust will melt, and the organics on the surface will probably react with water, each other and themselves. It'll be a wonderfully interesting organic chemistry mess. Most of the organic 'goo' will be floating on the surface of the water."

And yet there is a possibility that life already exists on Titan. Malaska says life could be based on different sets of molecules and interactions, and that microscopic alien organisms may already be swimming in seas of methane. Could this have a profound effect on our ability to colonise Titan in the event of a red giant? Would questions be raised over our chances of adapting and living alongside such alien life? Time will tell, and scientists will be working on those very answers.

"If we discover life on Titan, it would be incredibly huge," says Malaska. "It would be a fundamentally different type of life and would take our fundamental understanding of biological processes to a new level."

He claims that Titan has already changed how we think about geology: "Comparing and contrasting the geology of Earth and Titan is a powerful tool. With regards to geology, we talk about how lakes and rivers work and now have examples using both water and hydrocarbons, so we can understand the fundamental processes even though the materials, temperatures and gravity fields are totally different.

"Discovering life on Titan would likewise change our understanding of the fundamental processes of biology. It would extend our concept of the habitable zone where surface liquids exist to a different temperature range and set of surface conditions. It would also tell us that there may be even wilder and weirder temperature and chemical regimes that we can start to think about."

But even if life exists, emerges or travels to Titan, one thing is certain: it won't be staying there forever. Any migration from Earth to Titan will always be temporary since it will start to get too close to the Sun. Once those 100 million years are up, liquid water on Titan will evaporate and the moon will suffer an incredible rise in heat.

But there is potential for a reprieve. The Sun will later contract and become a white dwarf that will burn for a billion years. This will place Titan back into the habitable zone. It means that any humans who escape to Titan and then suffer another setback as Titan is burned dry could – if they somehow hold out – have another opportunity for a long existence on Saturn's moon. As Anderson told her audience: "Maybe life has a real chance on Titan during those billion years." For the sake of the future of humanity, we sincerely hope it does.

Setting up home around Saturn

It may have similarities with Earth, but what would it really be like to live on Titan?

Red giant
Residents on Titan would see the Sun as a huge, vivid red ball.

What a view
From some areas of Titan, around a third of the view will be taken up by Saturn.

A lovely atmosphere
The thick atmosphere is 1,000 kilometres (621 miles) high, compared to 100 kilometres (62 miles) on Earth, making deliveries easy.

Indoor life
People living on Titan would live indoors where the environment can be better regulated.

Suiting up
A spacesuit is only needed to breathe and keep warm. Walking will be like trekking through pillows.

Lots of lakes
Lakes on Titan could be used for transport and, by using dams and gates, power generation.

© Adrian Mann

"THAT'S ONE SMALL STEP FOR A MAN, ONE GIANT LEAP FOR MANKIND..."

Discover the missions that paved the way for Apollo 11's historic journey, the astronauts who followed in Neil Armstrong's first footsteps, and the brave crews that risked their lives in the name of exploration and space science.

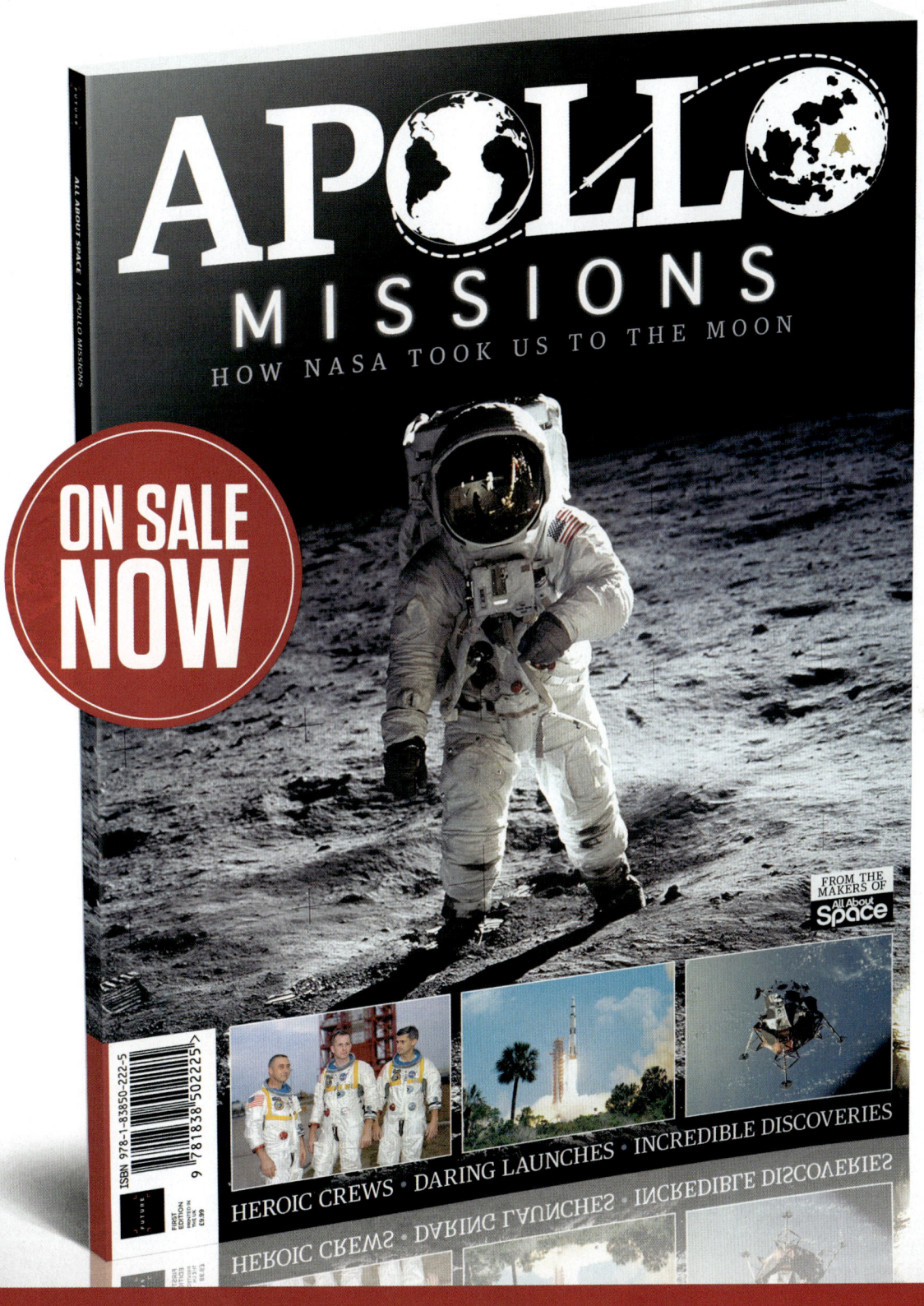

ON SALE NOW

Ordering is easy. Go online at:
magazinesdirect.com
Or get it from selected supermarkets & newsagents

Other Solar System phenomena

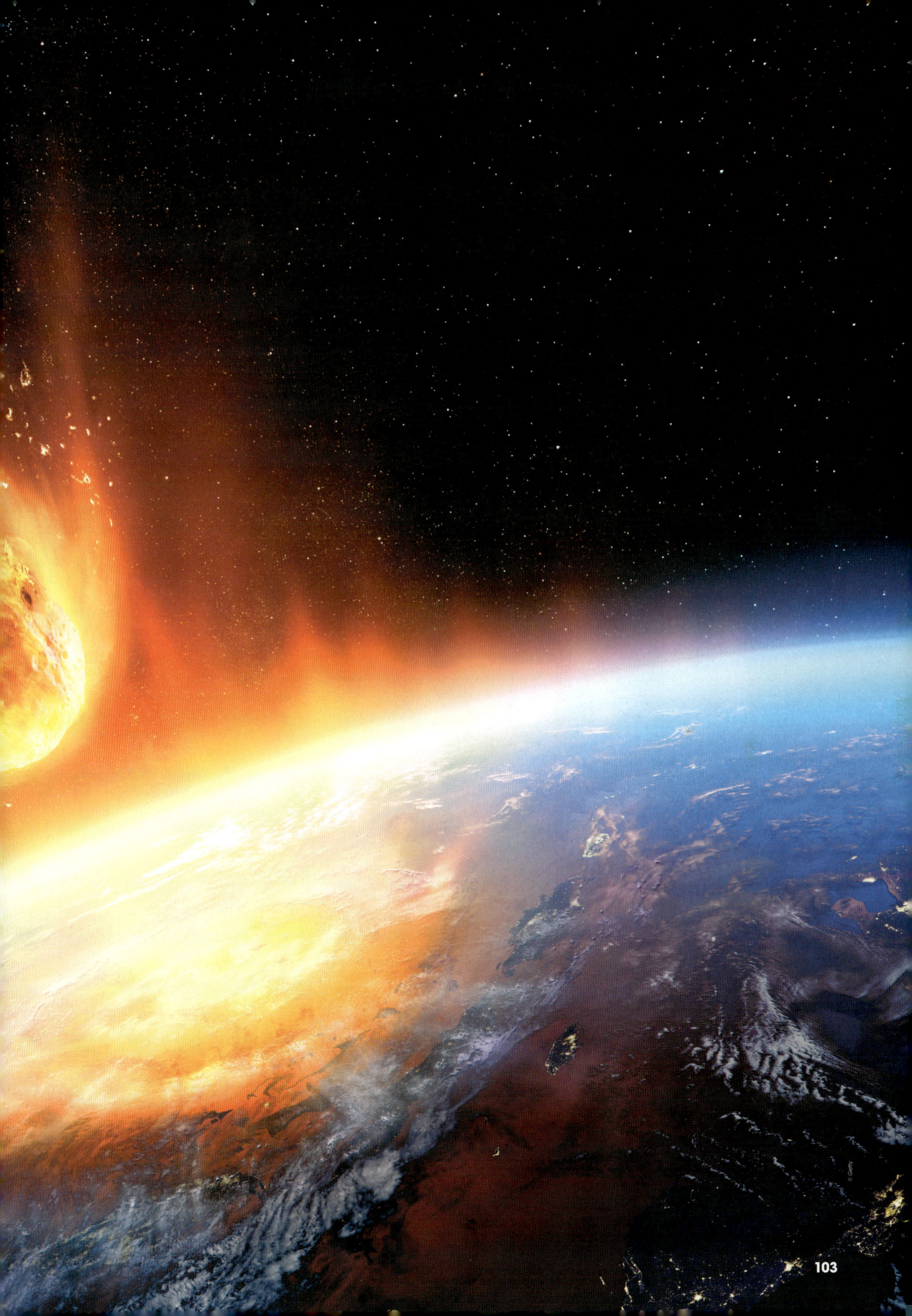

Space volcanoes

From Venus to Mars and the moons around far-flung planets, volcanoes have helped shape the bodies of our Solar System

On 5 March 1979, Voyager 1 made its closest approach to Jupiter. What it discovered astounded navigators at the Jet Propulsion Laboratory in California. Not least of all astronomer Linda Morabito, who had been analysing an image taken by the spacecraft and saw a puzzling feature that turned out to be a volcanic plume off the limb of Io. It was 270 kilometres (170 miles) tall, spewing sulphur into the airless sky with great ferocity. This volcano came to be known as Pele, after the Hawaiian fire goddess, and its discovery was hugely significant: it was the first time that an erupting volcano had been found anywhere other than Earth.

It wasn't the first time that alien volcanoes had caught the imagination. Missions to the Moon had uncovered basalt samples some 3.3 billion years old, and Apollo 15 landed close to Hadley Rille, an immense groove on the Moon 1.5 kilometres (0.9 miles) wide and 300 metres (984 feet) deep. This groove likely

SPACE VOLCANOES

UNDERSTANDING THE SOLAR SYSTEM

originated as a lava tube whose roof collapsed. The unmanned Mariner 9 highlighted a varied Martian terrain in 1977 which had huge volcanoes, including the mammoth Olympic Mons. Yet these discoveries were all completely extinct.

Io proved to be a swirl of colours thanks to a thin atmosphere laden with sulphur, and was showing signs of being the most geologically active body in the Solar System: more than 150 active volcanoes – of more than 400 in total – have been discovered there. Moons Enceladus and Triton also have active volcanoes. Venus, too, as well as the Jovian moon Europa. "Although we have volcanoes on Earth, you have to study somewhere different to understand the big picture," says Dr Rosaly Lopes of NASA's Jet Propulsion Laboratory.

Lopes became interested in volcanology during her studies, becoming particularly hooked when Mount Etna exploded and her volcanology professor didn't show up. "I thought it was really exciting to work on something where you had to rush off like that," she says. Following her graduation she worked on Galileo, a mission to Jupiter that launched in 1989. She studied infrared data from Io that allowed her to detect the heat from the volcanoes. Between 1996 and 2001 she discovered 71 active volcanoes – more active volcanoes than anyone else. "Io has sulphur dioxide pretty much everywhere," Lopes explains. "So you have to detect either the heat, plume or a surface change to show that a volcano is active. The easiest way is to detect heat, so I would compare pixels of the images we received, looking for infrared hotspots that were different to the surroundings. We had limits of resolution and it was a lot of work, but it wasn't that hard."

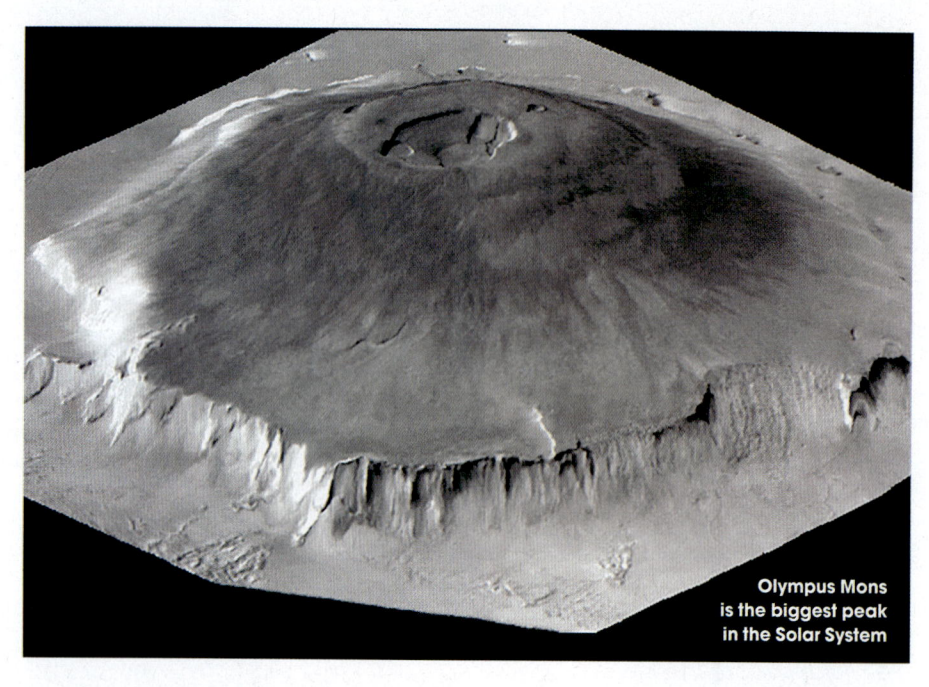

There are different types of volcanoes in the Solar System. Shield volcanoes are built up of fluid lava flows and they have broad, low-profile features. Composite volcanoes are conical, built up of ash, rock, dust and hot steam. Depending on their location in space, they either spew molten silicate rock magma or – as is the case beyond Mars – cold or frozen gases including water, ammonia or methane. Volcanic activity can be short or long-lived, continuing to spew for decades at a time. Io volcanoes stay active for very long periods of time. "They are much more powerful than the volcanoes on Earth," says Lopes. "When Voyager 1 flew past in 1979, about a dozen volcanoes were active. When Galileo visited, most of these volcanoes were still active and there were detections from the ground in-between. The New Horizons voyage to Pluto used Jupiter as a gravity assist. We did some observations of Io at that time, and some of the very same volcanoes were still active."

"There are some volcanoes on Earth that are always active on land. There are also volcanoes under the ocean that are harder to find. Certainly Io's volcanoes have the largest heat output and the largest calderas – craters formed by volcanic eruptions or the collapse of surface rock into a vacant magma chamber. Io is considered to be one of the most volcanically active bodies that we know of." The most unusual volcano on Io is Loki Patera. It's the most powerful and has the largest volcanic caldera in the Solar System. There is also potential evidence of a lava lake, usually a rare occurrence, but seemingly common on Io. "We want to know what is creating these lava lakes and how the eruption mechanisms work," says Lopes. "Loki Patera has some peculiar

Olympus Mons is the biggest peak in the Solar System

Io's volcanoes are very different to Earth's

SPACE VOLCANOES

Volcanoes of fire and ice

Volcanoes can blow hot and cold, but what's the difference between the two?

Two main types of volcanoes exist in space. The first and most familiar is the type that spews out molten rocks, typically at high temperatures of at least 700 degrees Celsius (1,292 degrees Fahrenheit). These exist on terrestrial planets and moons that are composed primarily of metals or silicate rocks – in our Solar System they tend to be closest to the Sun. When they erupt, magma leaves the volcano and reaches the surface; it then becomes known as lava. The volcanoes on Io, Venus and Mars – both active and extinct – are of this type. Our Moon has also had such volcanism. "Recently, volcanism that was around 100 million years old was discovered on the Moon," says Khan, pointing to NASA's Lunar Reconnaissance Orbiter, which showed the Moon's volcanic activity gradually slowed over time.

The second type of volcano is very different. Cryovolcanoes, which are colloquially known as ice volcanoes, still have a heated interior, but they spew water mixed with ammonia or methane rather than molten rock. Cryovolcanoes exist on icy moons such as Enceladus and Titan, which circle Saturn. "The water that comes up from the liquid ocean beneath the icy crust of these moons behaves very similarly to lava," says Lopes. "It's defined as volcanism because it's a process of bringing material to the surface."

1 Tidal heating
When there is tidal friction, the interior of the moon starts to become very hot.

2 Melting ice
It heats a pressurised H2O pocket that melts ices. Because the heat has to somehow escape, it begins to push upwards to the body's surface.

3 Eruption
When it breaks through, it sends a water vapour plume and ice particles into the air. The friction heats nitrogen, which builds pressure and erupts.

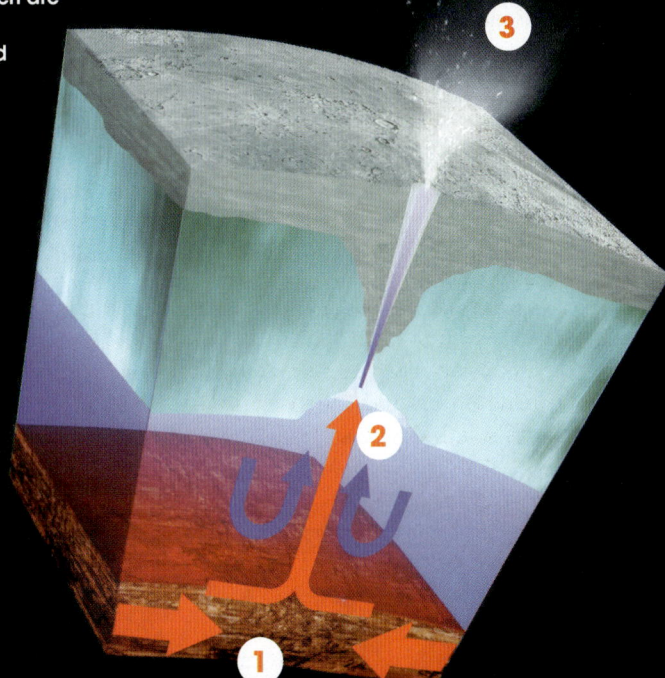

4 Main vent
A composite volcano's magma escapes through a large main vent at the top.

5 Shield volcano
Wider than composites, they have gentle, sloping sides; lava is able to flow out easily.

6 Multiple eruptions
Layers of hardened lava build up. A conical appearance appears as a result.

7 The crust
Magma breaks through the crust on its route up to the surface.

8 Secondary vents
The magma seeks other outlets and can also escape through secondary vents.

9 Shield magma
The magma chamber of a shield volcano is spread over a wider subsurface area.

10 Lava channels
Venus has lava channels. Its longest is 6,800 kilometres (4,200 miles) long.

11 Rising magma
As magma rises, pressure builds. The buildup is more intense under a composite volcano.

12 Chamber
Beneath the surface is molten liquid rock, pooled in what is called a magma chamber.

UNDERSTANDING THE SOLAR SYSTEM

Scaling space peaks

How does Mount Everest compare with some of the highest peaks in the Solar System?

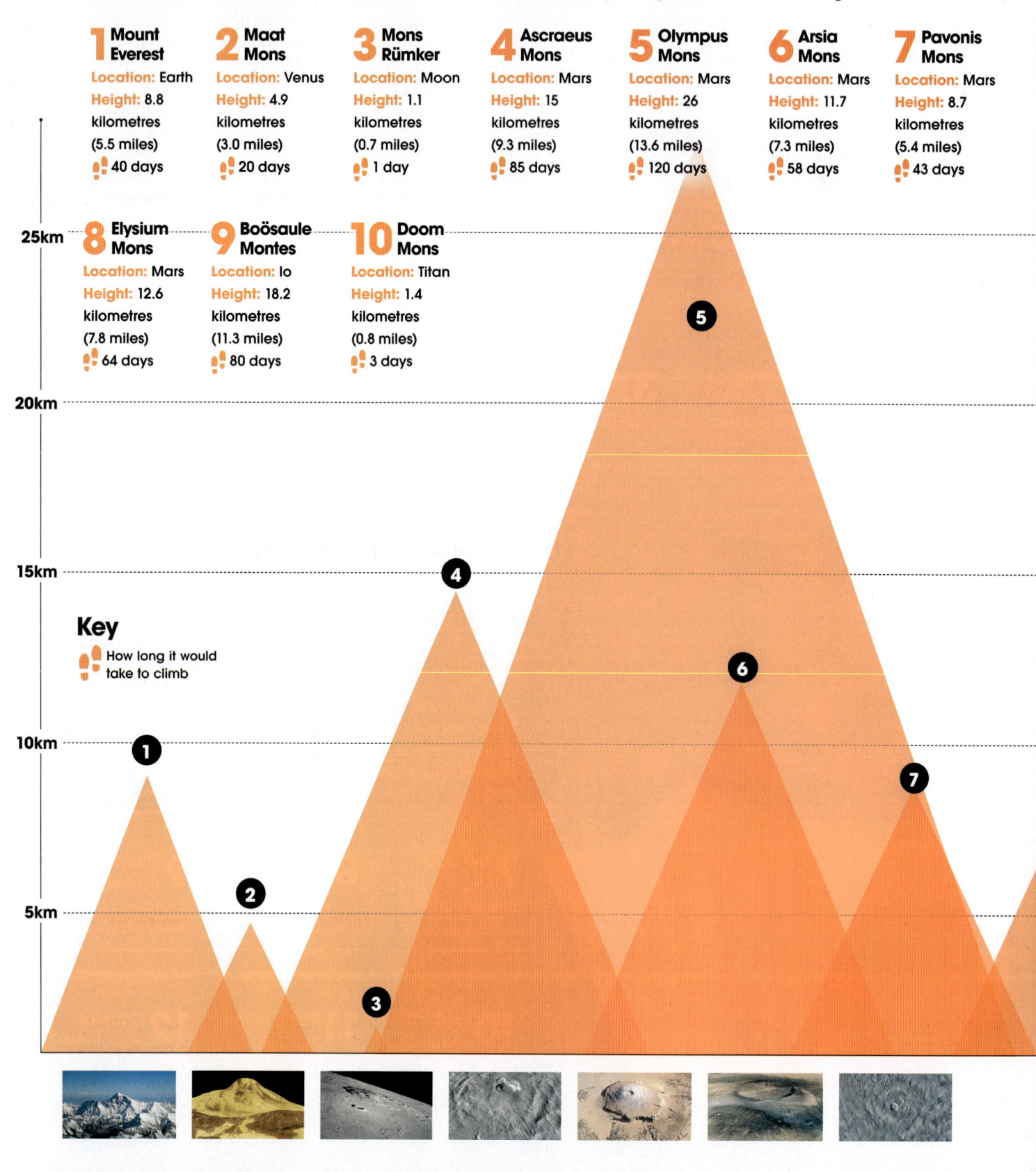

1 Mount Everest
Location: Earth
Height: 8.8 kilometres (5.5 miles)
40 days

2 Maat Mons
Location: Venus
Height: 4.9 kilometres (3.0 miles)
20 days

3 Mons Rümker
Location: Moon
Height: 1.1 kilometres (0.7 miles)
1 day

4 Ascraeus Mons
Location: Mars
Height: 15 kilometres (9.3 miles)
85 days

5 Olympus Mons
Location: Mars
Height: 26 kilometres (13.6 miles)
120 days

6 Arsia Mons
Location: Mars
Height: 11.7 kilometres (7.3 miles)
58 days

7 Pavonis Mons
Location: Mars
Height: 8.7 kilometres (5.4 miles)
43 days

8 Elysium Mons
Location: Mars
Height: 12.6 kilometres (7.8 miles)
64 days

9 Boösaule Montes
Location: Io
Height: 18.2 kilometres (11.3 miles)
80 days

10 Doom Mons
Location: Titan
Height: 1.4 kilometres (0.8 miles)
3 days

Key
How long it would take to climb

108

SPACE VOLCANOES

Dwarf planet Ceres has been found to have active cryovolcanism

patterns – an almost cyclical pattern of eruptions that we thought we understood, but then stopped."

How explosive a volcano is depends on its composition and the amount of gases dissolved in the magma. It's often been compared to shaking up a fizzy drink and opening it – if the drink has been allowed to go flat, it's likely to come out with less ferocity than one with its full soda potential. "In a volcano, you get these gases that are dissolved in the magma," says Lopes. "When magma rises towards the surface, the pressure lessens and the gases want to come out. If the lava is very viscous, or sticky, the gases cannot escape easily and eventually they will come out explosively. You can get what we call a Hawaiian-type eruption where you may have this beautiful lava oozing in fountains."

Interestingly, there are differences in the ways similar types of volcanoes behave according to where they are. On Earth there is a relatively thin crust that's divided into several plates gliding over the mantle. This is referred to as plate tectonics. The crusts slowly move, crash and slide into each other, propelled by the incredible heat simmering below them. When one plate is forced below another in a process called subduction, the magmas that come out in those places tend to be more viscous, forming explosive cone-shaped volcanoes. When they pull apart, more fluid basaltic lava comes out, creating shield volcanoes that erupt effusively, rather than violently. Plate tectonics are an alien concept on Io, where volcanic activity originates from the tidal forces associated with its planetary host, Jupiter. Io was not expected to have volcanoes since it is a small body and should have cooled a long time ago, like Earth's Moon. But as it rotates around Jupiter it is effectively squeezed as it gets closer to the planet in its orbit then moves away. Its surface is constantly being bent and flexed, creating the necessary heat for volcanism. It's like taking a ball of wax and massaging it, making it hotter and hotter inside.

"It's unusual because Io is around the size of Earth's Moon and it should have lost a lot of its primordial heat, just like the Moon has. It should have cooled down," says Dr Lopes. "But Io is in a peculiar orbit and it has the gravitational pull of Jupiter. At the same time, the other satellites further away from Io are also experiencing this pull. The constant tug-of-war causes friction, creating heat and ensuring the interior of Io remains very molten. That is what drives volcanism."

As volcanoes erupt on Io, it affects the entire Jovian system. The plumes, says Dr Michael Khan, who works in the mission analysis office of the European Space Operations Centre at the European Space Agency (ESA), produce a ring of charged material around Jupiter and "create a very nasty environment" as around 2 trillion watts of power is generated. "All of the stuff gets electrically charged. If you want to fly a spacecraft there, it gets hit by the charged particles and all of the electronics fry. It's not a nice thing to happen."

But the differences between volcanoes on different planets go beyond the causes of activity. Even patterns of eruption can be dissimilar, as is the case on Venus. "Venus can go hundreds of millions of years with no activity," says Khan. "Then everything goes off at once and the surface is completely remodelled." Volcanoes cover around 90 per cent of Venus and its surface has been transformed by volcanic eruptions. According to Lopes, who documented planets in her book Alien Volcanoes, the Magellan spacecraft found Venus' volcanism to be young in geological terms. "There are about 1,000 volcanoes, but I haven't counted them," she jokes. "The surface is really volcanic, but we don't know that much about it because it's a very challenging environment. The

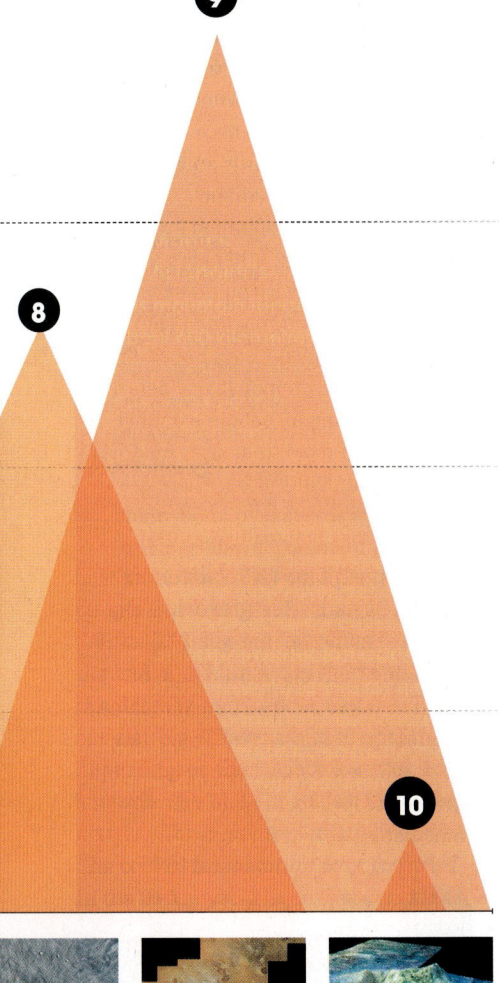

> "We want to know what is creating these lava lakes and how the eruption mechanisms work"
> **Rosaly Lopes**

UNDERSTANDING THE SOLAR SYSTEM

The robotic space probe MESSENGER spent four years orbiting Mercury, uncovering the planet's volcanic past. Hills, vents and long channels have been photographed. One of the volcanoes was thought to have erupted for a billion years

Most of the Moon is covered with hardened lava, made up of old basaltic flows. Some volcanic features may be less than 50 million years old

same is true of Io, as the environment is very radiation intensive."

Planetary volcanism can mimic that of the moons. For instance, Mars has dozens of volcanoes that are large and dominant, and this is believed to be due to lower surface gravity. A thicker crust and higher eruption rates allow lava to pile on top of lava, creating extra height and bulk. Olympus Mons is 25,000 metres (82,021 feet) high; not only is it three times as high as Mount Everest, but its footprint would also cover the entirety of Germany. If it existed on Earth at that size and weight, it would break through the crust and go right to the mantle. Mars also has limited plate movement, meaning the lava buildup only has certain areas in which it can break through. The surface of Mars is effectively two large tectonic plates that have been rubbing against each other.

"Volcanoes are so different," says Lopes. "While we have instrumentation that can give us geophysical measurements and an idea of what's underneath a volcano, there's a lot that we surprisingly don't know. There may be peculiar conditions that will make volcanoes erupt in a certain way. You only have to look at Mount Saint Helens in Washington, which erupted in 1980. Scientists were expecting the blast to go up, but it went sideways. A lot of people weren't evacuated because the scientists weren't modelling for it to go this way."

Mars may also have geological structures that are referred to as mud volcanoes. They are similar to the geysers in Iceland, spewing dirt from beneath the ground. A region called Acidalia Planitia in the northern plains of Mars appears to have a fair few of these, and they are also found on Earth – there is a large concentration in Azerbaijan and the adjacent Caspian Sea. "It may explain the plumes of methane in the atmosphere of Mars," says Khan, who notes that the drastic differences and completely unpredictable natures of many volcanoes make for some very intriguing observations. "It tells us how differently planets can evolve even though they were created at the same time and relatively close to each other," he says. "If you look at the exoplanets around other millions and billions of stars, how this enormous variety of different geologies can exist is a lesson that the Solar System is teaching us. It also shows that the conditions for life may exist in situations where you wouldn't think it is possible."

By way of explanation he points to Europa, another of Jupiter's geologically active moons. Like Io, it has been deformed by tides as it orbits the giant planet, releasing heat through rock-and-ice friction. Hubble revealed the icy moon spouting water in 2013, showing that the hypothesis of an underground ocean was probably correct. "Volcanic vents locally heat up the water, and eruptions happen," explains Khan. "Volcanism could be an enabler for life on Europa because it has warmth, nutritious minerals and water, which are the basic ingredients."

But just as some astronomers are looking for signs of life, others are keen to discover volcanic activity on other bodies. "It's why it's important to spread your studies to other planets," says Lopes. "You never know what you may learn. Before we started studying Io, people wouldn't have imagined a moon that size could have active volcanism, but it does. There are certainly things we want to know about volcanism on Io, the big mystery being the composition of the lava. Studying this would help us understand lava from the early Earth." Better equipment is helping enormously. "There have been a lot of advances in telescope instrumentation and techniques," says Lopes. "We are getting to the point where we may have smaller and cheaper telescopes that can observe these volcanic bodies very frequently."

David Crookes
Science and technology journalist
David has been reporting on space, science and technology for many years, has contributed to many books and is a producer for BBC Radio 5 Live.

EVERYTHING YOU NEED TO KNOW ABOUT A UNIVERSAL ENIGMA

Whether they're supermassive, primordial or double, black holes are a mystery. It's time to strip away some of puzzle as we head inside black holes to find out how they work and what really happens on the event horizon.

ON SALE NOW

Ordering is easy. Go online at:
F U T U R E magazinesdirect.com
Or get it from selected supermarkets & newsagents

UNDERSTANDING THE SOLAR SYSTEM

A megatsunami swept over Mars after a massive asteroid hit the Red Planet

The impact left a huge crater behind

A Martian megatsunami – a giant killer wave that may have reached more than 80 storeys tall – may have raced across the Red Planet after a cosmic impact similar to the one that likely ended Earth's age of dinosaurs. Although the surface of Mars is now cold and dry, a great deal of evidence suggests that an ocean's worth of water covered the Red Planet billions of years ago. Previous research found signs that two meteor strikes might have triggered a pair of megatsunamis about 3.4 billion years ago. The older tsunami inundated about 800,000 square kilometres (309,000 square miles), while the more recent one drowned a region of about 1 million square kilometres (386,000 square miles).

A 2019 study found what may have been ground zero for the younger megatsunami – Lomonosov crater, a 120-kilometre (75-mile) wide hole in the ground in the icy plains of the Martian arctic. Its large size suggests that the cosmic impact that dug the hole itself was big, similar in scale to the one from a ten-kilometre (six-mile) wide asteroid that struck near what is now the town of Chicxulub in Mexico 66 million years ago, triggering a mass extinction that killed off 75 per cent of Earth's species, including all dinosaurs except birds. A new study has found what may be the origin point of the older megatsunami – 111-kilometre (69-mile) wide Pohl crater, which the International Astronomical Union named after science-fiction grandmaster Frederik Pohl in August.

The scientists focused on the landing site of NASA's Viking 1, the first spacecraft to operate successfully on the Martian surface. Viking 1 touched down in 1976 in Chryse Planitia, a smooth circular plain in the northern equatorial region of Mars. The probe landed near the endpoint of a giant channel, Maja Valles, carved out by an ancient catastrophic flood, the first time scientists identified an extraterrestrial landscape carved by a river. Unexpectedly, instead of discovering the kind of flood-related features scientists had expected of the site, such as streamlined islands worn smooth by flowing water, they found a boulder-strewn plain.

Now the researchers suggest these boulders may be debris from a megatsunami, the giant wave carrying pulverised rock away from the site of the cosmic impact. "The marine floor would have been tossed up in the air, feeding the wave with sediments and probably aiding the development of a catastrophic debris flow front," said lead scientist Alexis Rodriguez, a planetary scientist at the Planetary Science Institute in Arizona.

The scientists analysed maps of the Martian surface created by combining images from previous missions to the planet. This helped them identify Pohl, which is located about 900 kilometres (560 miles) from Viking 1's landing site within a region of the Martian northern lowlands. "The northern plains of Mars comprise an enormous basin where, about 3.4 billion years ago, an ocean formed and subsequently froze," Rodriguez said. "The ocean is considered to have formed due to catastrophic floods released from aquifers, so my initial approach to looking for a megatsunami-triggering impact was to look for a crater beneath the ocean's frozen residue and above the channels that discharged the ocean-forming floods." Pohl was the only crater the scientists found that met this criterion, he noted.

The researchers simulated cosmic impacts on this region to see what type of impact might have created Pohl. Their findings suggest that Viking 1's landing site is "part of a megatsunami deposit emplaced about 3.4 billion years ago," Rodriguez said. Then the scientists used simulations to understand how a crater with similar dimensions to Pohl might have originated. If an asteroid encountered strong ground resistance, it would have needed to be about 5.6 miles

> **"The northern plains comprise an enormous basin where an ocean formed and froze"**
> Alexis Rodriguez

112

MARTIAN MEGATSUNAMI

(nine kilometres) wide, with the impact unleashing energy equivalent to 13 million megatonnes of TNT; if the asteroid met weak ground resistance, it might have been only three kilometres (1.8 miles) across, releasing the energy of 500,000 megatonnes of TNT. In comparison, the most powerful nuclear bomb ever tested, Russia's Tsar Bomba, had the strength of 57 megatonnes of TNT.

Both simulated impacts generated a megatsunami that reached as far as 1,500 kilometres (930 miles) from the impact site – more than enough to reach Viking 1's landing site. The massive wave might have initially stretched about 500 metres (1,640 feet) high and measured about 250 metres (820 feet) tall on land. Those statistics would make the Pohl impact similar to that of Chicxulub. Prior work has suggested that impact struck about 200 metres (650 feet) below sea level, formed a crater about 100 kilometres (62 miles) wide and triggered a tsunami about 200 metres (650 feet) high on land.

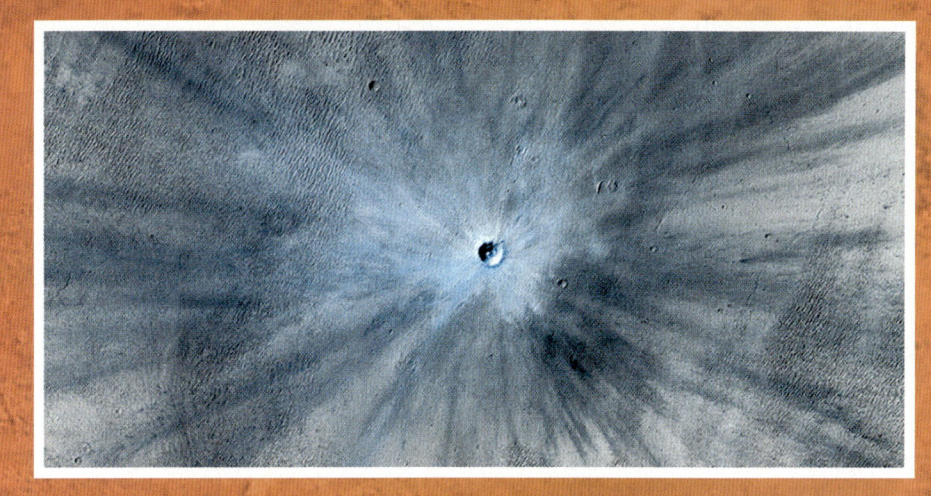

The researchers want to further investigate how the ancient Martian ocean might have changed between the two megatsunamis to see what potential biological effects that change may have had. "Right after its formation, the crater would have generated submarine hydrothermal systems lasting tens of thousands of years, providing energy and nutrient-rich environments," Rodriguez said.

A view from NASA's Mars Reconnaissance Orbiter of a crater on Mars that formed in 2012

The Red Planet
Explore Mars' fascinating surface features

113

UNDERSTANDING THE SOLAR SYSTEM

Comets, asteroids & meteor showers

Racing through space and crashing into Earth's atmosphere, we discover the space rocks that litter our Solar System

COMETS, ASTEROIDS & METEOR SHOWERS

Every 133 years, Comet Swift-Tuttle makes its return to the inner Solar System. It last made an appearance in 1992. Each time it nears the Sun, this speeding ball of ice, rock and dust grows a tail that deposits a glittering trail in its wake. Every year our planet moves through this, causing the dust particles left behind to come crashing through the atmosphere. Most of them are tiny, but as they burn up 100 kilometres (60 miles) above our heads they leave a bright streak of light. We call this a meteor, or 'shooting star'. The space between the planets and around Earth's orbit is full of dust, so every night there will be one or two random meteors. When Earth travels through the cloudy trail of dust left by a comet such as Swift-Tuttle, there are so many meteors that it's described as a meteor shower. If you've seen one, you'll know that meteor showers are among the most spectacular sights in the night sky.

There are many meteor showers each year, some better than others. The dust left by Swift-Tuttle forms the Perseid meteor shower, which runs from 17 July to 24 August each year. Each meteor shower has a peak, a time when the shooting stars occur in their greatest number. For the Perseids this occurs on 12 August. Either side of this peak, the number of meteors drops off – imagine the trail left by the comet beginning to spread out. The peak coincides with the densest part of the trail, and the most active meteor showers can produce more than 100 shooting stars per hour.

Other great meteor showers include the Quadrantids in January, which peak on the 3rd; the Lyrids between 16 and 25 April; the Orionids that peak on 21 October; the Leonids that are at their maximum on 17 November and the Geminids, which are at their best on 14 December. The names of the meteor showers come from the constellations in which they appear to streak from – this is the direction in which Earth is moving through the dust trails. For example, the Perseids streak across the sky from their 'radiant' in Perseus, the Leonids from Leo and the Geminids from Gemini.

The Geminids are unique among meteor showers. All the rest are produced by dust from comets, but the Geminids are

> **"If you've seen one, you'll know that meteor showers are among the most spectacular sights in the night sky"**

An antennae array under a meteor-filled night sky

115

UNDERSTANDING THE SOLAR SYSTEM

produced by dust left by an asteroid, 3200 Phaethon. This goes to show that sometimes the lines between asteroids and comets can be blurred. Astronomers have even witnessed some asteroids in the asteroid belt acting like comets by growing a tail.

Counting meteors is scientifically important. Differences in the speed, direction, brightness and colour of meteors can tell us a lot about the nature of the object that produced them. For example, Geminid meteors tend to move more slowly than meteors left by comets, and they burn up at a much lower altitude, about 38 kilometres (24 miles) above ground.

Occasionally, a meteor entering the atmosphere is a little larger than the rest. Rather than just leaving the thin streak of a shooting star, they are bigger and brighter, sometimes even brighter than Venus. These are fireballs that are burning up lower in the atmosphere. The brightest are called bolides – if you're lucky enough to see one, you might even see flaming chunks breaking off. Meteors around 20 to 30 metres (66 to 98 feet) in size will explosively fragment in the atmosphere, causing an airburst like the dramatic event that occurred over the Russian city of Chelyabinsk in 2013. This exploded 30 kilometres (18 miles) above the ground, and the shock wave shattered windows, damaged roofs and sent 1,500 people to hospital with cuts from flying glass.

The biggest meteors can actually reach the ground before they completely disintegrate. We call these meteorites, and over 60,000 of these have been found on Earth. Most are chunks of rock smaller than your hand, and the most common place to find them is in the white, icy landscape of Antarctica or a barren desert. Here the charred, black rocks stand out like a sore thumb. Not all meteorites come from asteroids – a handful come from the Moon and Mars.

NASA and the Japan Aerospace Exploration Agency (JAXA) have both launched asteroid sample-return missions, where a craft visits a celestial body and brings back material for examination on Earth. But meteorites are naturally occurring samples, bringing pieces of other celestial bodies to Earth. They're not pristine, having been blasted into space – probably as the result of an impact – before being burned in the atmosphere and landing on the ground on Earth. But they can tell us a great deal about the geology and chemistry of planets and asteroids. There has been speculation that some of the 277 meteorites from Mars contain evidence for life in the form of microbial fossils. This was a claim made by NASA scientists in the 1990s after examining a meteorite from Mars called ALH 84001, which was found in the Allan Hills region of Antarctica in 1984. Unfortunately, most scientists are now convinced that the microscopic features are not fossils at all, or if they are then they are fossils of microbes from Earth that contaminated the meteorite while it lay on the ice in Antarctica.

Meteorites can also tell scientists plenty about the dawn of the Solar System and the birth of Earth. This is because many meteorites represent debris left over from the distant era when the planets were forming 4.5 billion years ago. They are broadly split into three types: stony meteorites, iron meteorites and a mixture of the two. Stony meteorites, especially a specific type called chondrites, make up the vast majority of meteorites and are the same type of rock that built planets like Earth. They are very primitive, having never really melted, and so they preserve the chemical building blocks of the planet-forming disc that surrounded the young Sun.

The other type of stony meteorites are called achondrites, and these have melted

A meteor shower occurs when Earth dances into a path of debris left behind by a comet or asteroid

Asteroid types
The differences between four main categories of these space rocks explained

253 Mathilde
C-type
Carbonaceous

They're so dark due to their carbon-black surfaces that even the largest require a telescope to detect. They consist mostly of clay and silicate rocks and account for more than 75 per cent of all asteroids. Most of these ancient space rocks orbit in the outermost regions of the asteroid belt.

433 Eros
S-type
Stony

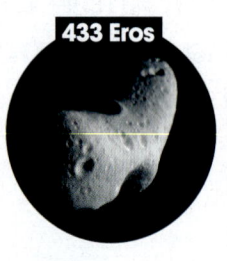

This class of asteroids orbit the inner asteroid belt and are primarily composed of stony materials, metallic nickel-iron as well as iron and magnesium silicates. S-types are the second most common asteroids and are also among the brightest – some larger examples, such as 7 Iris, can be spotted with binoculars.

21 Lutetia
M-type
Metallic

M-types are pure metal, or mixtures of metal and small amounts of stone, and have originated from the cores of planetary bodies that have been broken apart by impacts. Most are metallic, comprising largely of nickel-iron, and they are found in the middle region of the asteroid belt.

4 Vesta
V-type
Vestoid

V-types have similar surface properties to 4 Vesta, one of the largest asteroids in the Solar System. They're not so different in composition to S-type, also made from stony iron and chondrites, but they contain higher levels of silicon-aluminium oxides called pyroxenes. They are a reddish colour.

COMETS, ASTEROIDS & METEOR SHOWERS

Naming space rocks

Depending on their size and what they're made of, space rocks take on new names

1 Meteor shower
These occur at the same time every year, when Earth passes through a region that has a large concentration of debris shed from either a comet or an asteroid. From our location on Earth, meteors appear to originate from the same location year after year.

2 Comet
These bodies are made of ice, rock, dust and frozen gases. Comets have a nucleus and show off a brilliant tail when they get closer to the Sun. As they disintegrate, some comets leave a trail of solid debris.

3 Meteoroid
A small rocky or metallic body that races through space, meteoroids are quite a lot smaller than their larger cousins, asteroids. Lumps of space rock that are even smaller than meteoroids are classified as micrometeoroids, or space dust.

4 Asteroid
Any large lump of space rock over one metre (3.3 feet) in size is classed as an asteroid. They often pass our planet and are found most commonly in the asteroid belt between Mars and Jupiter.

5 Meteor
The streak of light that's thrown out by a meteoroid or asteroid as it enters the atmosphere at high speed. The brightness comes as the rock rubs against air particles to make friction, heating the meteor.

6 Fireball
This is another term for a very bright meteor. If you ever see a fireball streaking through the night sky, you'll quickly notice its bright-white to orange hue outshines that of the brightest planet in the sky, Venus.

7 Bolide
Similar to fireballs, but in this instance their brightness is likened to that of a full Moon and even brighter. Bolides often explode in the atmosphere.

8 Meteorite
If a piece of another celestial body survives its passage through the atmosphere and touches down on the ground, we call this piece of space rock a meteorite. They can weigh in at anything from a few grams up to dozens of tonnes.

UNDERSTANDING THE SOLAR SYSTEM

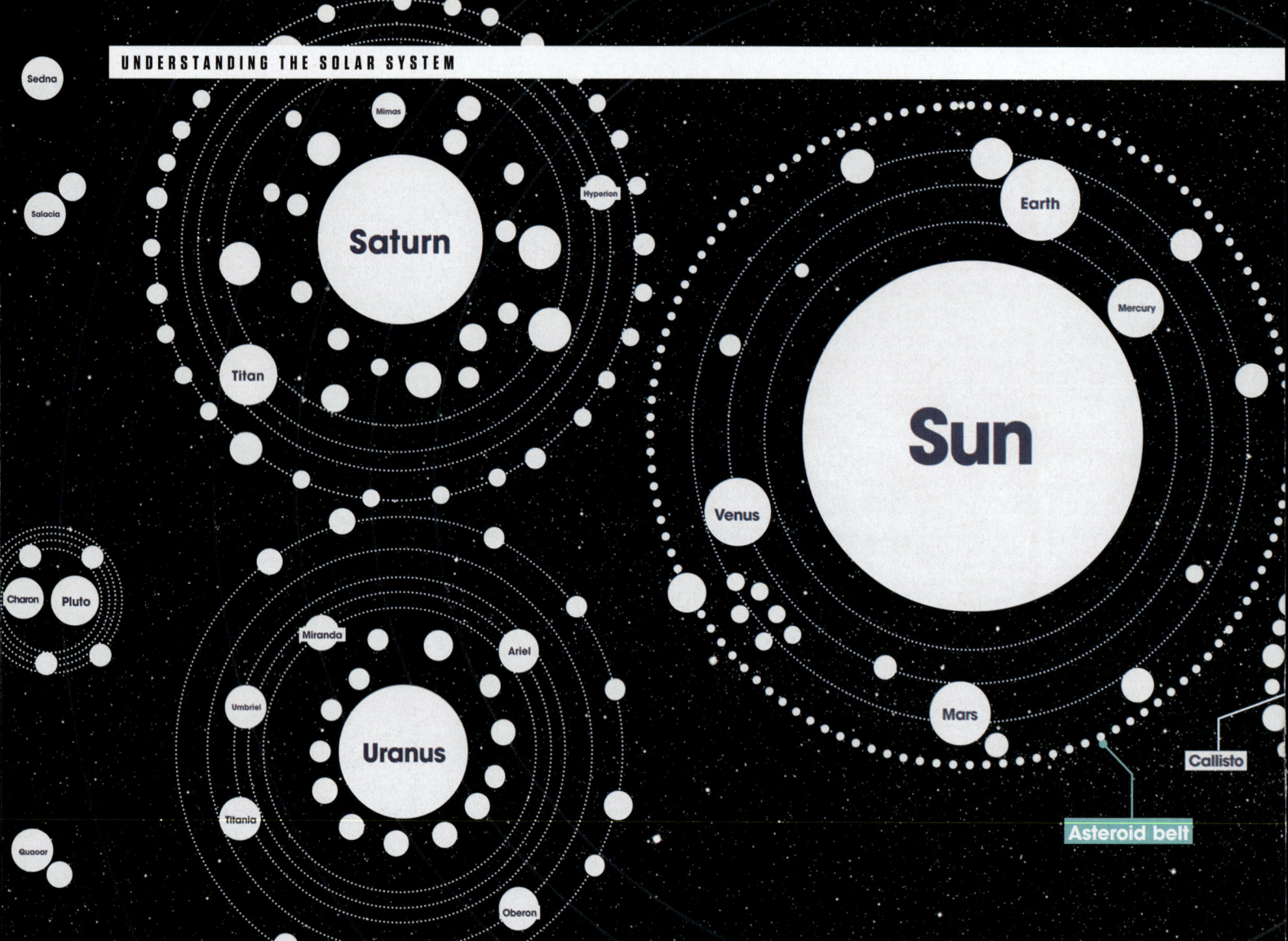

Celestial bodies

Discover some of the many dwarf planets and moons orbiting in our Solar System

Makemake
Found around 4.2 billion miles from the Sun, just outside the orbit of Neptune, this dwarf planet is the second-brightest object in the Kuiper Belt – the first being Pluto. Its discovery in 2005 prompted the International Astronomical Union to form a new classification of celestial bodies, called dwarf planets.

Haumea
Haumea sits in the Kuiper Belt and is one of the fastest rotating large objects in the Solar System. A single day on Haumea is equivalent to four hours on Earth, but due to its proximity to the Sun, one Haumean year is equal to 285 Earth years. This oval-shaped dwarf plane has two moons: Namaka and Hi'iaka.

Io
Io is one of the most volcanically active bodies in the Solar System. There are hundreds of volcanoes on the moon's surface, each of them spewing lava dozens of miles high, along with lakes of molten silicate. It's thought that Jupiter's intense gravitational pull is the reason for Io's explosive nature.

Callisto
Callisto has a circumference of 9,410 miles, which is almost as big as Mercury. Not only is this moon impressively large, it has a salty secret deep below its icy surface. Discovered in 1610, it wasn't until the 1990s that scientists proposed the moon has a subsurface ocean about 155 miles below its surface.

Europa
Another of Jupiter's many moons, Europa is one of the oddest. With a surface temperature of around -160 degrees Celsius, this frozen satellite bears strange streaks. These markings are thought to be cracks in the moon's icy surface, caused by the tidal forces of an ocean deep beneath it.

Titan
Although the structure of Titan remains unclear, scientists think its core is made of rock around 2,500 miles in diameter, surrounded by water ice. This satellite has a dense atmosphere, which gives it its yellow hue. The composition of this atmosphere is primarily nitrogen and some methane.

118

COMETS, ASTEROIDS & METEOR SHOWERS

Near-Earth objects are monitored by many organisations to prevent disaster

Mimas
Often called the Death Star moon for its similarity to the space station in Star Wars, Mimas is one of Saturn's smallest moons. Its iconic impact crater, named Herschel after English astronomer William Herschel, who discovered Mimas in 1789, spans 80 miles and reaches 3.5 miles high at its peak.

Hyperion
Not all moons are spherical. Some, like Saturn's sponge-like moon Hyperion, are irregular and filled with deep caverns. With a lower density than water, this moon is made up of water ice and frozen methane or carbon dioxide. Hyperion's appearance is thought to be the result of its distance from Saturn.

– either in the impacts that blasted them off their original asteroid or when they were buried deep inside a large asteroid where conditions were hot. Achondrites are special because they tell us about the chemical conditions within large asteroids and the protoplanets similar to those that eventually became the true planets.

Iron meteorites also come from the cores of protoplanets, because that's where all the iron sank to when they formed. Iron meteorites are incredibly hard and dense, but only five per cent of all meteorites are of this variety. There are two types of stony-iron meteorites: pallasites and mesosiderites. Pallasites are recognisable thanks to their large crystals of a green mineral called olivine. Mesosiderites are more of a jumble of rock and metal, made when two asteroids collide in space, with the impact fusing different materials together.

On rare occasions, a really big meteor will enter Earth's atmosphere. These are sometimes big enough to blow out large craters or explode over towns and cities, causing harm. Over 100 years before the Chelyabinsk meteorite, a similar airburst flattened 80 million trees in Tunguska, a remote region in Siberia.

These events greatly worry scientists, who fear that one day an asteroid will hit us that could destroy a city or send so much dust into the air it would block out the Sun and end life on Earth. It's a big concern to Queen founder and astrophysicist Dr Brian May, who lends his support to the Asteroid Day event to raise awareness about this threat: "30 June was the anniversary of Tunguska in 1908. Not a huge object, but it exploded before it hit the ground, which flattened the trees for hundreds of miles around. Now that is a city destroyer – the force of a thousand atom bombs."

To forewarn us, NASA's Spaceguard program has found over 90 per cent of asteroids larger than a kilometre (0.62 miles) that come close to Earth. These are the real killers, like the asteroid that wiped out the dinosaurs 65 million years ago. However, there are still millions of undiscovered asteroids out there smaller than 100 metres (328 feet) that could still do serious damage.

Astroid-surveying missions remind us that we still have much to do to protect ourselves from asteroids. Comets can also be a danger, but because there are fewer of them they pose less of a risk. Instead Earth is more likely to fly through their tails so we can see spectacular meteor showers in the sky. From shooting stars to fireballs, meteors and meteorites, their origins are all the same, just on vastly different scales.

Colin Stuart
Science author and speaker
Colin holds a degree in astrophysics, has written over 17 books on space and has an asteroid named in his honour: 15347 Colinstuart.

UNDERSTANDING THE SOLAR SYSTEM

Where are the biggest craters?

Earth's largest craters are big, but these are some of the biggest in the Solar System

The largest confirmed crater in the Solar System is the one that lies beneath Utopia Planitia, a huge plain in the northern hemisphere of Mars. With a diameter of 3,300 kilometres (2,050 miles), this huge and shallow basin must have been created by the force of a wide impactor and would have had a devastating effect on Mars at the time. But the Utopia impact happened roughly 4 billion years ago, and later geological activity has done a lot to hide it. The basin was only confirmed in 2001 using satellite maps of Martian topography. Far more easily identified as an impact basin is Hellas Planitia, an oval depression in the Red Planet's southern hemisphere some 2,300 kilometres (1,429 miles) across. What's more, these same maps revealed that a vast swathe of the northern hemisphere of Mars is notably depressed and flat compared to the rest of the planet.

Impact craters in the outer Solar System tend to be significantly smaller, with the largest only a few hundred kilometres across. This may be partly because the impacts that bombarded the larger asteroids and moons were genuinely less powerful, but it's also for the simple reason that smaller objects can only withstand lesser impacts without shattering apart completely. The asteroid Vesta, for example, is misshapen by the vast Rheasilvia crater, whose diameter of 505 kilometres (314 miles) is a full 90 per cent of Vesta's ideal spherical size. Several of Saturn's icy moons display craters with 30 to 40 per cent of their diameter, and their creation must have come close to breaking them up.

Scale
One centimetre is equal to 200 kilometres

Mars

Mars

Iapetus

1 Turgis
The largest impact crater in the outer Solar System lies on Saturn's icy moon Iapetus. At 580 kilometres (360 miles) across, its diameter is 40 per cent that of Iapetus itself.

2 Utopia Planitia
The largest confirmed impact basin on Mars creates a huge bay on the edge of the southern highland terrain and contains the landing site of NASA's Viking 2.

3 Mars' north polar basin
What might be the Solar System's largest impact basin creates a huge elliptical depression around the Martian north pole that's partially obscured by volcanic rises.

4 South Pole-Aitken basin
This vast crater on the lunar far side extends down to our satellite's south pole. Despite its size, it has never been flooded with volcanic lava, so it retains its original depth.

WHERE ARE THE BIGGEST CRATERS?

6 The Moon

7 Mercury

8 Mercury

4 The Moon

9 Mars

5 The Moon

Vesta

10

Expert: Robin Hague
Robin is a science writer, focusing on space and physics. He is head of launch at Skyrora, coordinating launch opportunities for Skyrora's vehicles.

5 Procellarum basin
This enormous suspected basin is filled by Oceanus Procellarum, a lunar ocean that dominates the Moon's western hemisphere as seen from Earth.

6 Imbrium basin
The Imbrium basin is the second-largest lunar impact crater and the only one to have been precisely dated. Rock samples suggest it formed nearly 4 billion years ago.

7 Caloris basin
Mercury's biggest crater. Shock waves from its formation spread either side of the planet, creating a jumble of chaotic terrain where they met up again on the far side of Mercury.

8 Rembrandt
This sharply defined basin is the second largest on Mercury. It's roughly 3.9 billion years old and was discovered by NASA's MESSENGER probe during a 2008 flyby.

9 Hellas Planitia
Mars' most obvious impact crater, the Hellas basin is a deep depression in the planet's southern highlands. Red dust blown from around the planet accumulates here.

10 Rheasilvia
This enormous dent in the south pole of Vesta changes the shape of the entire asteroid. Rheasilvia is the largest impact crater relative to its parent body in the entire Solar System.

121

UNDERSTANDING THE SOLAR SYSTEM

Alien Storms
Discover incredible weather on other worlds and what causes it

ALIEN STORMS

UNDERSTANDING THE SOLAR SYSTEM

Our angry, stormy Sun

Is our Sun becoming more deadly?

We know our Sun as a brilliantly bright sphere that rises in the east and sets in the west each day. That's a simple way to describe it; what really occurs on its surface is far from the impression it gives as it hangs, almost calmly, in the daytime sky.

While going near the Sun would be suicide, with the searing heat and penetrating radiation combining to fry you alive in your spacesuit, technology has revealed this star to be an angry, bubbling cauldron of solar activity.

First up are solar flares – bursts of radiation from the sudden release of magnetic energy from active regions on the Sun's surface, the photosphere. These regions are centred on sunspots, which are tangled knots of magnetic fields. The flares release as much as a sixth of the total amount of energy that the Sun releases every second, with much of it in X-rays or ultraviolet light. The energy of a flare can drive a cloud of charged particles to escape the solar corona in a coronal mass ejection (CME). The CME becomes a giant cloud of plasma hurtling through space and, when CMEs are pointed towards Earth, they cause solar storms.

The Sun's surface is a hive of constant and violent solar activity

When a CME strikes the Earth's magnetosphere, it overloads the system and becomes a geomagnetic storm. Earth's magnetosphere is compressed to breaking point with charged particles flooding the magnetic field lines that loop down on to the magnetic poles of the planet. The particles excite atmospheric gases (mainly oxygen and nitrogen), causing them to glow in eerie shimmering curtains of light – the aurora borealis (northern lights) and the aurora australis (southern lights). Oxygen gas glows green, while nitrogen glows purplish-red – the two primary colours seen in auroras. Usually, low-level solar wind activity means that the 'auroral arc' is kept, in the northern hemisphere, to the Arctic Circle, but the power of a geomagnetic storm can see the auroral arc extend to more southerly latitudes over Britain and western Europe, as far south as Spain or even, on very rare occasions, Florida in the United States. The most severe solar storm on record was the Carrington event of 1859, when auroras lit up the skies as far south as the tropics and telegraph wires began to short, sparking electricity.

Those telegraph wires remind us that auroras are only the pretty side of a geomagnetic storm. Although they are not directly harmful to people on the ground, a storm instigated by a powerful CME can destroy our technology. Satellites can short-circuit, knocking out communications. Astronauts must shelter from radiation in a shielded room on board the International Space Station. On the ground, power lines can be swamped by raw current from the CME plasma – in 1989, a solar storm caused a nine-hour blackout in Quebec in Canada. In our modern world, the

Solar wind current

1 Surface of the Sun
As the solar wind rises it simplifies until it consists of two polarities separated by the line of the heliospheric current sheet.

2 Corona
In the corona, the solar wind begins to draw out the heliospheric current sheet into space, extending the Sun's atmosphere out into the rest of the Solar System.

3 Rotation
As the Sun rotates, it causes the heliospheric current sheet to become twisted.

4 Jupiter
Material in the heliospheric current sheet takes three weeks to reach Jupiter. The sheet extends into the Kuiper belt, where Voyager is exploring.

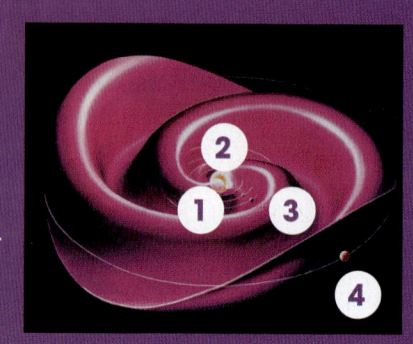

ALIEN STORMS

How solar radiation storms are classified

Minor
50 per 11-year solar cycle
A minor solar radiation storm causes minimal impact on high-frequency (HF) radio in the polar regions, but otherwise causes no damaging effects.

Moderate
25 per cycle
A moderate storm affects navigation at the polar caps and may, in rare instances, cause problems in satellites, but poses no threat to humans.

Strong
10 per cycle
During a strong storm, astronauts are advised to seek shelter, while satellites could lose power and instrument usage. HF radio will degrade at the poles.

Severe
3 per cycle
Astronauts and passengers on planes may be exposed to radiation, while satellites could experience orientation problems. HF radio blacks out at the poles.

Extreme
Less than 1 per cycle
Astronauts and aeroplane passengers exposed to high radiation. Satellites may be rendered useless. HF communications black out in polar regions.

The aurora borealis (northern lights) and aurora australis (southern lights) can be seen in the northern and southern hemispheres of our planet

nightmare scenario is that a powerful enough solar storm could stop everything working, wiping computers, crashing the internet, knocking out power systems and disrupting communications. The world would be sent into technological, social and economic chaos.

We're most vulnerable to solar storms at solar maximum, the point in the Sun's 11-year cycle of activity when our nearest star is most active. Solar flares happen all the time, and CMEs strike Earth frequently, but rarely are they as powerful as the solar activity that plunged Quebec into darkness. However, scientists are currently unable to predict solar activity or when the next big CME will be.

All of this takes place in the Sun's heliosphere, which is the extent of its magnetic influence throughout the Solar System, where the solar wind still blows. The heliosphere goes out past the orbit of Pluto. The Voyager 1 spacecraft is 118

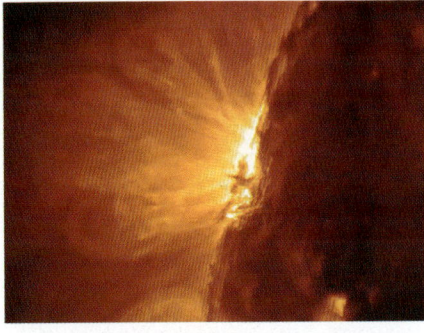

A solar prominence is an eruption of hydrogen gas from the Sun's surface

times further from the Sun than Earth is, yet it has still to leave the heliosphere. CMEs disperse and lose power the deeper they get into the Solar System. However, solar activity can still have an effect, even on the edge of the heliosphere. Both Voyager 1 and 2 have experienced the heliosphere swelling and shrinking on gusts of the solar wind that inflate the Solar System's magnetic bubble.

How can a spacecraft help?
How ESA's bold proposal would help predict incoming solar weather

1 The Sun
As the Sun rotates, a coronal mass ejection (CME) or solar flare can come into view.

2 L1
L1 sees solar weather at the same time as Earth, so doesn't give the same advance warning.

3 Incoming particles
L5 could give us more accurate information on particles heading for Earth.

4 Advanced warning
Ships at L5 would see the Sun up to five days before it rotates into view of Earth.

5 L5
L5 is about 60° behind Earth. A probe here can judge the speed of solar ejections heading to Earth.

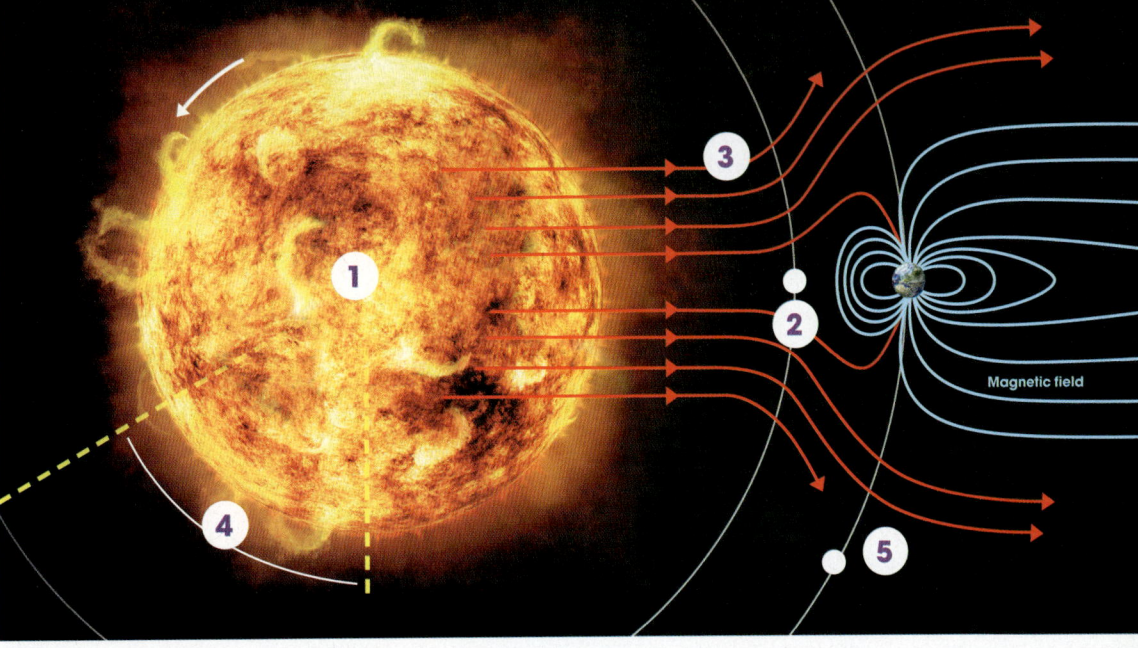

Magnetic field

125

UNDERSTANDING THE SOLAR SYSTEM

Dust storms that cover the planet

How solar heating drives up immense storms on the Red Planet

Now, this is really bad weather – a dust storm that doesn't just cover an area, or a hemisphere, but the entire planet. During summer in the Red Planet's southern hemisphere, when Mars is at its closest point to the Sun, solar heating can drive immense storms that blow up red dust and can obscure the surface for months. In 1971, when Mariner 9 arrived on Mars, it found the whole planet hidden under a veil of dust, with only the volcano Olympus Mons visible. More recently, the Mars Exploration Rovers Spirit and Opportunity would struggle to survive in dust storms as the Sun's light was blocked and their solar panels covered by a coating of dust.

On Earth, moisture arms swirling storms, but on Mars there is only dust. Normally most of the dust is on the ground, but some is found in the atmosphere, where it scatters sunlight and makes the sky appear pinky-red. When Mars is at its hottest – still cold enough to freeze water – the atmospheric dust can absorb the energy of the sunlight, which causes warm pockets of air to rapidly move towards colder, low-pressure regions, generating winds of up to 45 metres per second (162 kilometres per hour or 100 miles per hour) that begin to pick up dust particles from the ground, adding to the atmospheric dust content and increasing heating, pushing the winds harder and faster until the atmosphere is filled by dust.

And then, just as quickly, the storm can die down. Perhaps by blocking the sunlight, the surface of Mars grows cooler, allowing some of the dust to begin sinking out of the atmosphere. Not all dust storms swallow the entire planet – some are more localised. However, were you to be on the surface during a dust storm, other than the sky darkening and a fine coating of dust settling over you, the atmosphere is so thin that you'd barely notice the wind or the scouring dust.

Snaking its way across Mars's surface, this dust devil is powered by solar heating just like the dust plumes found on Earth

How do dust storms form?

1 Heating up the atmosphere
The absence of clouds or water means radiation cannot be reflected back into space, and the thin atmosphere close to Mars's surface becomes hotter than the atmosphere above it.

2 Picking up the dust
As the atmosphere is heated, dust is lifted into the air and, after absorbing more sunlight, the dust warms up the atmosphere further, propelling more dust into the air.

3 The storm begins
The change in temperature creates winds, swirling at great speeds of 96 to 193km/h (60 to 120mph), capable of dominating the entire planet.

ALIEN STORMS

Kicking up dust

1 Desert dust
The dust storms that frequently rise from the cold deserts of Mars, sometimes rage across the entire Martian globe, which crackle and snap with electricity.

2 Electrifying dust
It is possible that dust particles could be electrified in Martian dust storms when they rub against each other as they are carried by the winds, transferring positive (+) and negative (-) electric charges.

3 Strong swirls
Electric fields generated by dust are thought to be strong enough to break apart carbon dioxide and water molecules in the Martian atmosphere, recombining to make reactive chemicals like hydrogen peroxide, which you'll find in bleach and ozone.

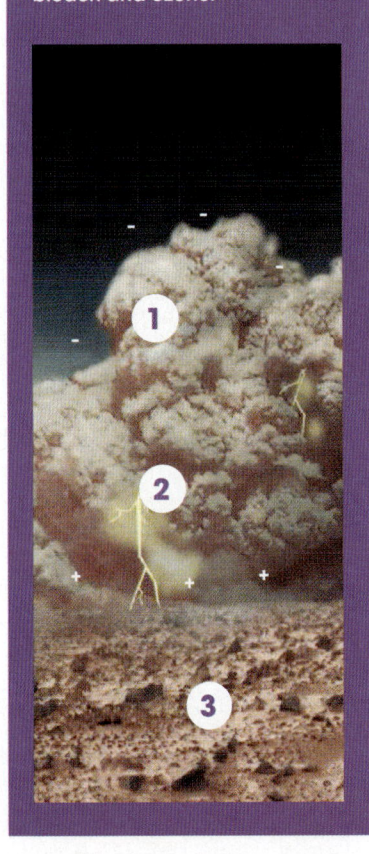

Hurricanes bigger than Earth

How did the Great Red Spot appear on Jupiter's surface?

Easily one of the most famous storms in the Solar System, Jupiter's Great Red Spot is so large that it is visible through many Earth-based telescopes.

The Great Red Spot is thought to have been in existence for at least 340 years. The oval red eye rotates anticlockwise due to the planet's crushing high pressure. Winds can reach over 400 kilometres per hour (250 miles per hour) around the spot. However, inside the storm they seem to be nearly non-existent. And that's not all; this complicated weather system has an average temperature of about -162 degrees Celsius (-260 degrees Fahrenheit).

At around eight kilometres (five miles) above the surrounding clouds and held in place by an eastward jet stream to its south and a strong westward jet flowing into its north, the Great Red Spot has travelled several times around Jupiter, but how did it appear on the gas giant's surface?

The answer is not clear, despite the efforts of scientists. However, experts do theorise that the storm is driven by an internal heat source, and it absorbs smaller storms that fall into its path. Another thing they know is that the Great Red Spot hasn't always been its current diameter. In 2004, astronomers noticed that it had around half the 40,000-kilometre (25,000-mile) diameter it had around 100 years before. If the Great Red Spot continues to downsize at this rate, it could morph from an oval shape into a more circular storm by 2040. You might think this well-known feature won't be sticking around for long as it becomes smaller, but experts believe it's here to stay since it is strongly powered by numerous other phenomena in the atmosphere.

Storms like these are not out of place on Jupiter, whose atmosphere is a zigzag pattern of 12 jet streams, with blemishes of warmer brown and cooler white ovals in the atmosphere owed to storms as young as a few hours or centuries old.

The science of the Great Red Spot

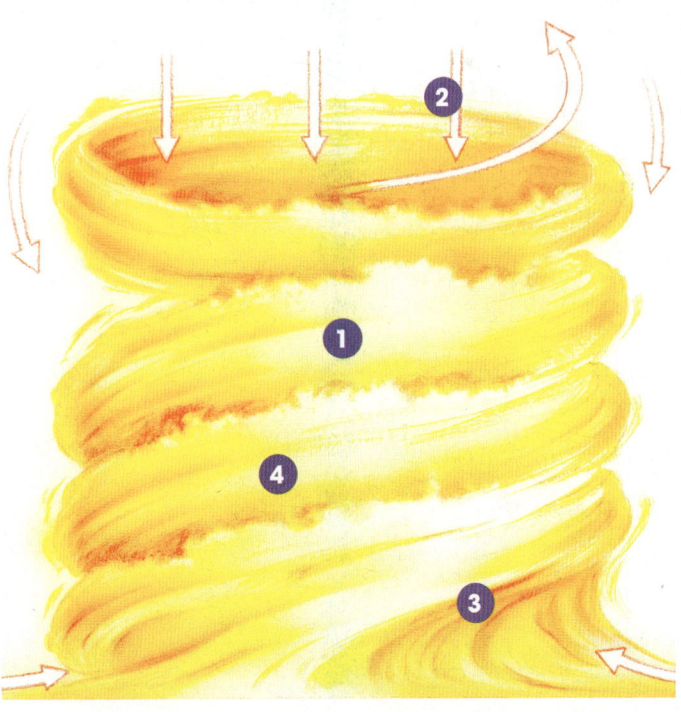

1 A constant twirl
Hot gases in the atmosphere constantly rise and fall.

2 Falling cool gas
Cooler gas falls through the atmosphere. Coriolis makes the area whirl, creating eddies.

3 Shifting and merging eddies
Eddies move around and merge, creating more powerful storms.

4 High wind speeds
Winds of the Great Red Spot can reach over 400km/h (250mph).

© NASA; Getty

UNDERSTANDING THE SOLAR SYSTEM

The violent polar vortex

How Saturn's polar storms create a warm spot on the planet's surface

On the outside, Saturn almost looks like a calm, bland world, but once in a while, huge storms flare up on the planet. From the short-lived Great White Spot of 1990 to the more recent storm of 2010, which grew into an atmospheric belt covering around 4 billion square kilometres (1.5 billion square miles), Saturn has proven to be a turbulent world. What's more, the storms on Saturn are the second fastest in the Solar System, after ice giant Neptune, peaking at an impressive 1,800 kilometres per hour (1,120 miles per hour) and blowing in an easterly direction.

Temperatures on Saturn are normally recorded at around -185 degrees Celsius (-300 degrees Fahrenheit), but near the giant swirling polar vortex – a persistent cyclone taking pride of place at the ringed planet's south pole – temperatures start to warm up, and while the climate doesn't reach high enough to get a suntan, this

Around once every Saturn year (roughly 30 Earth years), huge, storms work their way through the clouds of the northern hemisphere. The storm pictured here, which was imaged in 2011, is the longest storm to date, lasting roughly 200 days

1 Counterclockwise swirl
This storm angrily swirls in an anticlockwise direction, rotating with a period of nearly 11 hours.

2 Monstrous size
Not only is this storm violent, it is also argued to be an estimated 4,000km (2,500 miles) wide – roughly the distance between New York and Los Angeles!

3 Rolling cloud formation
The bubbling of frothy clouds sit at the centre of Saturn's famed northern vortex, a hexagonal-shaped feature characteristic of the planet's two poles.

4 Fast and furious
This swirling vortex, located above Saturn's north pole at the centre of a jet stream, whips around at a speed of 480km/h (300mph) and is believed to be at least 30 years old.

ALIEN STORMS

-122 degrees Celsius (-188 degrees Fahrenheit) vortex is the warmest spot on Saturn, with a powerful jet stream smashing its way through this terrifyingly fierce feature.

Saturn's north pole also has a giant storm of its own surrounded by a persistent hexagonal cloud pattern. Spotted in 1980 and 1981 during the Voyager 1 and Voyager 2 flybys, Saturn's hexagon, complete with six clear and fairly straight sides, is estimated to have a diameter wider than two Earths. The entire structure rotates almost every 11 hours.

Sighted more closely by NASA's Cassini spacecraft in 2009 as spring fell on the ringed giant's northern hemisphere, experts believe the storm could have been raging for 30 years, whipping around at over 480 kilometres per hour (300 miles per hour) in a counterclockwise direction and disturbing frothy white clouds in its wake.

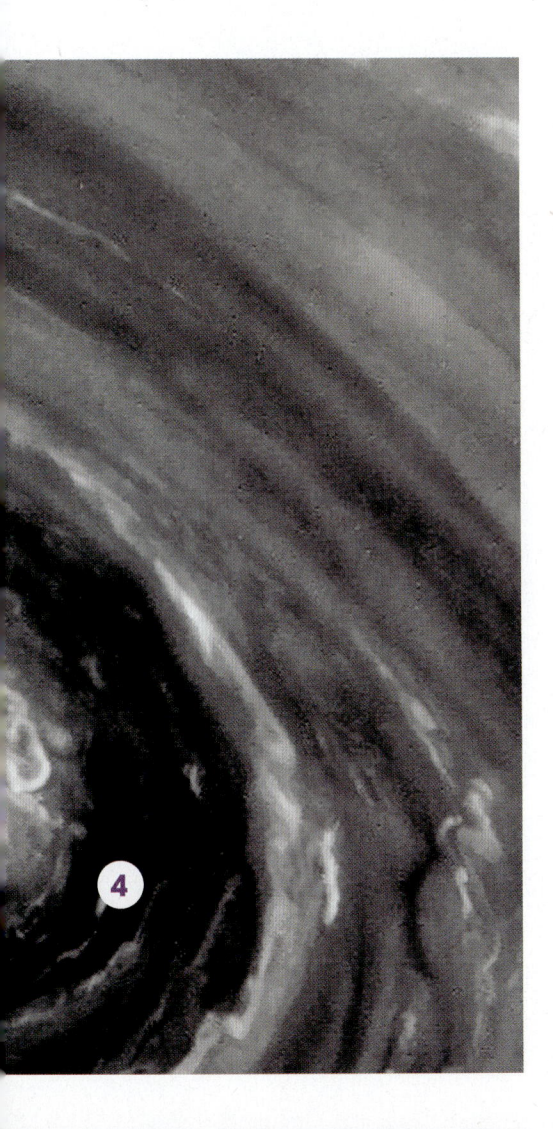

Deadly methane rain

Liquid methane rains down on the surface of Saturn's largest moon

With a surface pressure almost one and a half times that of Earth's, Titan's atmosphere is slightly more massive than our planet's overall, taking on an almost chokingly opaque haze of orange layers that block out any light that tries to penetrate the Saturnian moon's thick cover.

Titan is the only world, other than Earth, where liquid rains on a solid surface. However, rather than the water that we are used to falling from the skies above us, pooling into puddles and flowing as streams and rivers, this moon's rains fall as liquid methane – liquid hydrocarbons that add more fluid to the many lakes and oceans that already cover the surface. It is thanks to the moon's complex methane cycle, similar to the natural processes found on Earth, that this is possible.

Rain falls quite frequently on Earth. However, the same can't be said for some regions on Titan. Spring brings rain clouds and showers to Titan's desert with the moon only experiencing rainfall around once every 1,000 years on its arid equator. However, these rain showers certainly make up for the lack of activity by dumping tens of centimetres or even metres of methane rain on to the Titanian surface.

At the poles of the moon it's a completely different story, however. Methane rain falls much more frequently, replenishing the lakes of organic liquid covering the Titanian land.

> "Titan is the only world other than Earth where it rains on a solid surface"

UNDERSTANDING THE SOLAR SYSTEM

Winds at twice the speed of sound

Are Neptune's freezing temperatures responsible for the planet's incredible storms?

We've all witnessed strong winds here on Earth, from gusts that turn your umbrella inside out to tornadoes that rip up everything in their path. You might think these winds are a force to be reckoned with, but unless you've had a day floating around the gaseous atmosphere of ice giant Neptune, you haven't seen anything yet!

You might think that Neptune's distance from the Sun, which creates temperatures as low as -218 degrees Celsius (-360 degrees Fahrenheit), would mean a world frozen solid by the sub-zero climate with not much going on in terms of weather. However, you would be incorrect. The winds that race through its hydrogen, helium and ammonia-laden atmosphere can reach maximum speeds of around 2,400 kilometres per hour (1,500 miles per hour), making this dark horse probably the most violently stormy world in the Solar System, and making our most powerful winds look like light breezes.

Neptune's fastest storms take the form of dark spots, such as the anticyclonic Great Dark Spot in the planet's southern hemisphere and the Small Dark Spot further south – thought to be vortex structures due to their stable features that can persist for several months – as well as the white cloud group, Scooter.

So what causes these winds? Neptune might be extremely frosty, but astronomers think that the freezing temperatures might be responsible; decreasing friction in the gas giant to the point where there's no stopping those super-fast winds once they get going.

Delving into its layers of gas, we find another possibility pointing to just how these active storms came about as the temperature starts to rise. As things get more snug closer to the centre, the internal energy could be just what is driving the most violent storms that we've ever witnessed.

> "Neptune's fastest storms take the form of dark spots"

Long clouds on Neptune's surface are similar to cirrus clouds on Earth

Neptune's atmosphere

1 A stormy surface
Storms on the surface appear in the form of blemishes.

2 Great Dark Spot
This anticyclonic storm was replaced by a similar feature called the Northern Great Dark Spot.

3 Clouds and storms
The cyclonic storms are thought to occur in the troposphere at low altitudes compared to the brighter white clouds.

4 Small Dark Spot
This storm, also called The Wizard's Eye, is the second most violent storm on Neptune. Just like the Great Dark Spot, the Hubble Space Telescope found that this cyclone had disappeared in 1994.